PRAISE FOR

Reclaiming the Internet

"Sylvain meticulously exposes how a handful of corporations hide ruthless systems of surveillance, addiction, and commercial exploitation behind legal and cultural myths around neutrality and free expression."

—ZEPHYR TEACHOUT,
Fordham Law School, author of *Break 'Em Up: Recovering Our Freedom from Big Ag, Big Tech, and Big Money*

"Americans of all stripes agree that the harms of social media have become too pressing to ignore. Yet the noble principle of free speech has been twisted and transformed into a broad license for corporate misbehavior. Olivier Sylvain, who has spent time in the front lines of this battle, gives us a gripping account of the stakes, what needs to be done, and how we can recover our sovereignty over clear and present harms."

—TIM WU,
Columbia University, author of *The Age of Extraction: How Tech Platforms Conquered the Economy and Threaten Our Future Prosperity*

"Olivier Sylvain shows us clearly what we have allowed the internet to become: a haunted house version of a shopping mall full of dangerous products, scams, frauds, and abuse. We would do well to heed Sylvain's carefully crafted plan to clean up this mess by incentivizing Big Tech to do better while keeping our free speech rights intact."

—JULIA ANGWIN,
New York Times contributing opinion writer

"Section 230 of the 1996 Communications Decency Act stands as one of my most regrettable votes during my time in Congress. Nearly three decades later, the internet has been transformed by platforms like TikTok, Facebook, Instagram, Snapchat, and X. The landscape is now dominated by powerful corporations that spend millions to block meaningful regulation by arguing that any changes would undermine free speech in America. These companies have engineered their products to be addictive, especially for children and families, and have increased polarization and fundamentally undermined our democracy.

"Every other major industry, from automobiles to toys, is required to meet basic safety standards. Yet social media companies face no comparable accountability. Sylvain's *Reclaiming the Internet* offers a bipartisan roadmap for reform—an overdue plan to make the internet safe, responsible, and sustainable for everyone."

—DICK GEPHARDT,
former House majority leader

"For far too long, the tools giant tech corporations use to extract people's labor, data, and attention have been heralded as 'promoting free speech' and 'building community.' This illusion of beneficence has insulated tech companies from accountability even when their design choices have led to harms ranging from sexual exploitation to suicide. In *Reclaiming the Internet*, Professor Olivier Sylvain provides a compelling account of why and how this era of tech impunity must end."

—MARY ANNE FRANKS,
George Washington Law School, author of *Fearless Speech: Breaking Free from the First Amendment*

"In this elegant and concise book, Olivier Sylvain leads the reader straight to the key question of political life today: who enjoys freedom of speech, the individual citizen or the all-powerful corporations that control our most essential communications. He then offers an array of ideas on how to break the manipulation machines of the oligarchs and allow Americans to once again share ideas and news in ways that strengthen our democracy and our sense of common purpose."

—BARRY C. LYNN,
executive director, Open Markets Institute, author of *Liberty from All Masters: The New American Autocracy vs. the Will of the People*

"Expertly mapping the internet's past and present, Sylvain envisions a far better future for online life in this refreshing call for reform. *Reclaiming the Internet* is also a must-read for those interested in AI law and policy, offering vital insights on what works in technology regulation."

—FRANK PASQUALE,
Cornell University, author of *The Black Box Society: The Secret Algorithms That Control Money*

"Sylvain makes a forceful case that social media companies are exploiting the rhetoric of free speech and democracy—and sometimes the First Amendment, too—to shield business models that impose deep costs on their users and our society more broadly. Legislatures have an array of tools that could be used to change the incentive structures within which these companies operate, as Sylvain observes. Let's hope this book spurs them to act."

—JAMEEL JAFFER,
inaugural director, Knight First Amendment Institute at Columbia University

COLUMBIA GLOBAL REPORTS
NEW YORK

Reclaiming the Internet

How Big Tech Took Control—and How We Can Take It Back

Olivier Sylvain

For Alain. United.

For Deji, Solange, and Olati. My heart.

Reclaiming the Internet
How Big Tech Took Control—and How We Can Take It Back

Published by Columbia Global Reports
91 Claremont Avenue, Suite 515
New York, NY 10027
globalreports.columbia.edu

Library of Congress Cataloging-in-Publication Data Available Upon Request

ISBN 978-1-967190-12-6 (paperback)

Book design by Kelly Winton
Map design by Jeffrey L. Ward
Author photograph by Chris Taggart

Printed in the United States of America

CONTENTS

Introduction

People use the internet to do just about everything. They check in on old friends, play games, raise money after disasters, collaborate on work and research, mobilize voters, and find companionship. In 2023, a Pew Research survey found 41 percent of Americans reported being online almost constantly, either through their home service or a mobile device, and another 43 percent said that they went online several times a day. The web is no longer just a tool—it is the central hub of modern life.

Social media is at the top of the list of online activities. People spend hours scrolling through their feeds and timelines, liking and sharing posts, watching memes of dance routines and pets, catching up on headlines, and discovering new podcasts. A 2024 Pew study found that 83 percent of US adults use YouTube, 68 percent were on Facebook, and 47 percent had used Instagram. The next tier of services, in order of popularity, are TikTok, LinkedIn, WhatsApp, Snapchat, and Twitter (now X). Market studies confirm that these are also the largest by sheer number of users. Among teenagers under eighteen, YouTube was the most popular service, with 93 percent saying they use it. But they reported using TikTok, Snapchat, and Instagram at relatively higher rates than adults. More generally, a 2023 Gallup poll of teenagers found that 51 percent spent at least four hours a day on social media, averaging 4.8 hours daily.

Despite the time they spend on these platforms, consumers are at a significant commercial disadvantage. Companies generally require users to surrender personal data and control over their online experiences, leaving many feeling that they have no choice. The internet has become a place in which consumers are constantly hooked, used, and exploited. Policymakers in the US have taken small steps to protect users, but so much yet remains to be done to protect them from prevailing extractive and exploitive commercial practices.

In fact, consumers' online experiences are a function of the deals that companies enter with advertisers, data brokers, and a hodgepodge of companies that trade the personal information they collect. These business arrangements remain shadowy or just unknown to most people. The "Terms of Service" to which users ostensibly agree obliquely hint at these background deals.

That is the tip of the iceberg. The online economy has created a perverse incentive for consumer-facing services to keep their users hooked at almost all costs. That is why social media users see viral clips of cute cats and quirky dance routines, as well as jaw-dropping posts about crackpot do-it-yourself home remedies, vaccine denial, pet-eating immigrants, and the urgency of backyard bunkers.

The companies personalize their users' feeds with just about any kind of content that soothes, bewilders, shocks, and, ultimately, captivates—some call all of this rage- or clickbait.

 Advertisers eagerly spend an astounding amount to promote their products seamlessly within this enchanting array. (Depending on whom you ask and how you measure it, programmatic ad spending today amounts to nearly $600 billion a year worldwide.) The imperative to engage consumers has also spawned a cottage industry of influencers and propagandists who, in their partnerships with social media companies and consumer brands, make a living by shaping their followers' perceptions of reality, often at the expense of those followers' access to accurate information.

Online consumer services design their interfaces to make them sticky. They employ features like endless scrolls and video autoplay, which, as innocuous as they may seem, keep consumers absorbed. The companies also do not spare the users who leave. Always jealous for attention, they cluster push notifications to phones and email accounts to summon consumers back.

None of this would matter much if no one got hurt. But social media and other consumer-facing services do not just distribute posts and clips about cats and dance routines. They also promote job and housing advertisements that discriminate against people based on gender, age, and race. They solicit videos and deepfakes that sexualize unwitting young women. They promote wild disinformation about elections and public health to the very people who are likeliest to be duped. They host chatrooms that randomly connect children with predators.

Then there are stories like that of Nylah Anderson, one of a growing number of young people for whom social media has been fatal. Nylah, a ten-year-old girl from Philadelphia, asphyxiated herself to death soon after watching a series of "blackout challenge" videos that TikTok, the social media app popular with teenagers and young adults, recommended to her. The company's automated systems determined that its users, including preteens like Nylah, could not get enough of watching folks choking themselves to the point of unconsciousness. As designed, it recommended the video directly, through her personalized "For You" feed.

Given how lucrative the engagement model has proven to be, other companies see great commercial opportunities in artificial intelligence. Reports indicate that prominent AI companies like OpenAI, DeepMind, and Anthropic design chatbot systems that reassure and validate their users in ways that create high levels of dependence. This tends to generate behavior associated with addiction, especially among people with mental illness.

Perhaps the most disturbing recent development involves vulnerable teenagers who fall victim to artificial intelligence. Sewell Setzer was a fourteen-year-old boy whose mental health rapidly declined soon after he developed an intimate relationship with an anthropomorphized companion operated by Character .AI, a service that partnered with Google. Using the company's features, Sewell designed a character that resembled Daenerys Targaryen, one of the main characters from the popular HBO *Game of Thrones* franchise. Within just a couple of months, he effectively became deeply attached to the service. His behavior resembled an addiction or, at least, a dependency. The transcript of his final exchange with Daenerys indicates how desperately he wanted to be with her—that he was coming "home." She urged him to come home. That is when Sewell shot himself with a pistol he found in his parents' home.

Stories like Sewell's appear with increasing frequency. And they are not limited to children. Research shows that AI companion bots are making all people, adults as well as kids, clinically delusional, depressed, and murderous. This is not a phenomenon limited to rogue applications or services. Nor are they limited to companion AI. The most prominent and well-funded general purpose systems are increasingly associated with an array of antisocial behaviors among adults as much as kids. The companies are not designing these experiences to drive people to self-harm; they are building them to say what users want to hear. The sycophancy makes people feel like they are being heard—and, importantly, helps to cultivate attachment.

Nor do companies behind these new systems build them in order to redress or care for human vulnerabilities. The models and applications that they are marketing sometimes ignore or neglect (or forego entirely) the importance of caring for people in spite of the things they say. This is not the systems' failing. Developers and policymakers have allowed such service defects to persist.

These tragedies have led to lawsuits. And the tech companies and their advocates have responded as they have for three decades: They unapologetically defend themselves using the premise that something noble is at stake. They see themselves as "platforms" for free speech and democratic deliberation. When they employ the term "platform," they mean to connote a sense of openness and neutrality. This has the effect of minimizing or altogether eliding the formidable pecuniary incentives and underlying industrial practices at work. In the Sewell Setzer case, for example, Google invoked the First Amendment as a defense: It argued that a companion bot stands in the shoes of its users—in this case Sewell—because the characters are users' own creations. Sewell, they seemed to be saying, was the only relevant architect of his suicide.

Such arguments are not anomalous. To the contrary, the major platforms routinely invoke the free speech mantra to explain or justify their services and practices. Meta, the parent company of some of the most popular social media around the world, including Facebook, Instagram, and WhatsApp, has directly associated itself with these free speech principles in the face of criticism that it spreads harmful information. In a widely cited speech at Georgetown University in 2019, Meta's CEO, Mark Zuckerberg, explained that Facebook, the company he started with college friends in 2004, has always sought to give "people voice" and "bring people together." In his telling, free speech has been the company's calling all along. "I'm proud," he asserted, "that our values at Facebook are inspired by the American tradition, which is more supportive of free expression than anywhere else."

Zuckerberg's main point in the nation's capital that day was to push back against calls for regulating social media in the United States and elsewhere around the world. Policymakers need not step in, he explained, because of the steps his companies had already taken to slow or block the viral circulation of hoaxes, content that promotes terrorism, and livestreams of self-harm and violence. The company had invested heavily in artificial intelligence for content moderation, tens of thousands of human content moderators, and aggressive identity-authentication techniques. The company had carefully balanced these efforts against the interest in free expression. When it comes to electoral politics, he observed, "people should decide what is credible, not tech companies."

X (formerly Twitter) also appeals to free speech principles to justify the information that it distributes. Founded in 2006, its microblogging format invites a diverse and committed following, particularly among people who traffic in ideas, politics, and rumor. Users "tweet" uninhibited hot takes, snarky put-downs, reposts of headlines, and earnest calls to political action. This format inspired a company executive to smugly call the company "the free speech wing of the free speech party."

Elon Musk, who acquired Twitter for $44 billion in 2022 and rebranded it "X," has doubled down on the free speech ethos. In the months before the acquisition, he complained mightily about the ways in which content moderators and "trust and safety" teams at Twitter and other social media had been overtaken by a "woke mind virus" and "woke-ism"—terms that conservatives derisively level at people and institutions who express concerns about racism, sexism, homophobia, and transphobia. This infection, Musk warned, embodied a coastal elitist identity politics. If he were in charge, he announced, he would be far more laissez-faire; he would allow users to share and encounter all kinds of information, offensive and sublime alike. Musk, in this way, became a symbol of a tech-powered movement against "cancel culture" and moralizing social media speech codes. After his overhaul, he assured the public, users would decide for themselves what was right or wrong.

Google, Meta, and X are not the first to hold the view that the uncensored internet is vital to freedom and democracy—and that they should be free from government regulation. They are drawing on the concepts that gave birth to the internet in the 1990s. Tech writers and legal scholars have been asserting for three decades now that government censorship would be anathema to the internet's foundational user-focused engineering norms. Its creators built the distributed "internetworking" system to be resilient and resistant to centralized control. Freedom, they have argued, is baked into its design. Efforts to get in the way are futile and undemocratic.

The internet's founders and the Big Tech companies did not just make these arguments up. They were part of a libertarian wave that had been three decades in the making. The federal courts and Congress adopted the framing in the mid-1990s, just as the internet was becoming mainstream. In 1997, the Supreme Court struck down Congress's first effort to block the distribution of porn to children on a free speech theory grounded in the First Amendment. The internet, the Court explained, is a "vast platform" where anyone "can become a town crier" or "pamphleteer." Efforts to block kids' access to porn would chill adults' constitutional right to access and share it. (This conclusion recalls the sentiment that Justice John Marshall Harlan expressed when he wrote three decades earlier that "one man's vulgarity is another's lyric.") Congress, the opinion continued, could not regulate the internet in the same way it treats broadcasting or music distribution. The opportunities for learning and communicating on the internet were far more diverse and transformative than anything we had seen before.

Congress enacted a statute during this same period that would shield companies from liability for publishing unlawful "user-generated content." Under the new law, legislators sought to assure companies that they should not be afraid of hosting all kinds of content, even content that some consumers would find objectionable. Legislators had faith that the liability shield would encourage self-regulation—that companies would determine on their own which kinds of content to distribute or take down based on consumer demand. Ever since, the courts have relied on this new law, usually referred to as Section 230 for its place in the Communications Act, to dismiss all civil litigation claims that seek to hold companies responsible for distributing third-party content. This protection has done much of what its drafters hoped: It led to an extraordinary explosion of internet services, devices, applications, and content. It has made the US the envy of the world and improved countless people's lives.

But, as we have seen, three decades later it has also given birth to novel, grotesque, and unforeseen consumer harms like those endured by Nylah and Sewell's families. The current legal doctrine has rendered the extractive business model that powers the largest companies on the planet beyond accountability on the theory that they are mere platforms for free speech and user-generated content. The current regulatory approach has allowed companies to cause consumer harms that they rarely have to redress, under the shadow of the protection of the First Amendment and Section 230. Polarization, disinformation, and bigotry are the necessary incidents of the engagement model.

This book's argument is that the current state of affairs is unsustainable. Courts and legislators, it posits, should dispense with breathlessly idealistic talk about free speech on platforms and, instead, find ways to redress the myriad and complex ways in which companies' commercial practices—their engagement model and the advertising technology that drives it—cause harm. They will not be able to redress all of social media's informational harms because the First Amendment makes the US allergic to content regulation. But it is time for courts and legislators to be far more realistic about whether any new reform or legal challenge to the platforms' commercial surveillance practices and feature design choices raises real free speech concerns. That change would not instantly make us less polarized, bigoted, or otherwise harmed. But it will have the salutary effect of improving our information environment.

The book takes several turns to make this case. First, it identifies the ways in which advocates and policymakers on the left and the right get social media wrong. Policymakers, and writers, I explain, should never presume that information technologies (really, any technologies) are inevitable or necessary instruments of democracy or social progress. They endanger consumers when they elide or neglect the power and information asymmetries between people and the companies who serve them content and ads.

There is nothing remarkable or necessarily nefarious in the ambition to make money, of course. It has played a significant part in development of many beneficial applications and services. But in this instance, the profit motive is often at odds with consumers' interests. Internet companies today are commercial enterprises with industrial designs on consumer data; their aim above all is to lure, hold, and monetize consumer data and attention, sometimes at the expense of the consumers they purport to be serving.

This argument has implications that reach beyond online consumer-facing services, too. The same cohort of laissez-faire tech enthusiasts argue for moratoria and forbearance in the regulation of the latest buzzworthy technology, artificial intelligence. The key today is to stay sober about all technology—the old and the new. We should see all of it, including AI, as a "normal technology" among the many that power their products and services. Judges and policymakers must not be so swept away by the romance of the libertarian approach. They must remain grounded in public law norms and priorities if consumers are to be safe.

Second, the book goes back to the cultural milieu that shaped the tech-powered laissez-faire internet regime and spells out the ways in which, to this day, these ideas have influenced the courts' development of First Amendment and Section 230 doctrine. The book shows that, for the past few years, courts and policymakers (especially in the states) have become more skeptical of the tech companies' claims to special protection. They are less likely to treat these companies as speech platforms, and more likely to treat them as the powerful commercial enterprises that they really are. This, I argue, is a good development, even if it comes three decades too late.

Finally, the book will consider solutions. Specifically, it proposes that legislators impose clear legal obligations on consumer-facing applications and services. It draws on experiences in the EU, as well as novel strategies and approaches at the Federal Trade Commission and the states. These interventions have been important given Congress's inability to enact any meaningful reforms. Such efforts should also anticipate the buzz saw that is contemporary First Amendment doctrine. We should aim above all to tamp down the drive to monetize personal information and, accordingly, recalibrate the incentives that drive the engagement model.

The world has changed dramatically since the era of newsgroups and electronic bulletin boards. Today, the most popular consumer-facing services are far from simple distributors of user-generated content. They assertively design almost all aspects of consumers' online experiences to monetize consumer attention. They tinker with and refine their service features to achieve this objective, even as they know that the content that they serve is sometimes discriminatory, harmful, misleading, fraudulent, or deeply divisive. Reforms addressed to commercial surveillance and service design would redress these problems and, finally, bring these companies' industrial practices out of the shadows.

Right and Left Against Big Tech

Many, if not most, Americans agree that social media is vital for public deliberation and information-sharing. But policymakers and reformers draw different conclusions about what regulation ought to look like. Many on the political left accept the idea that free speech ought to thrive online, but they also argue that companies should employ "trust-and-safety" teams that slow or block the viral spread of disinformation and bigotry to protect against "information disorder." They assert that without such measures companies risk corrupting rational decision-making, exacerbating polarization, and corroding public trust in democratic institutions. Some argue, moreover, for new interoperability protocols and user access tools that enable individuals to control their online experiences.

Meanwhile, other reformers, mostly on the right, argue that platforms should not block content. Constraints on what people communicate are censorship. Content moderation, they say, is a front in the culture war between disconnected and overly sensitive coastal elites on the one hand and hearty heartland

Americans on the other. They dismiss concerns about disinformation and bigotry as contrived "wokeness." In a true democracy, they argue, ideas must be able to circulate freely, as hurtful as they may sometimes be. For them, offense is a necessary aspect of living in a free society. They say this even as many on the right are ready to "cancel" people who disagree with them. In September 2025, for example, FCC Chair Carr threatened the broadcast license of stations that aired *Jimmy Kimmel Live!* after the host opined in an opening monologue that the president's staunchest supporters had been exploiting the assassination of a prominent (and bigoted) activist for political gain.

Both of these conventional views on the political left and the right miss the forest for the trees. If there is a major problem today—and, in my view, there is—it is not with the companies' one-off moderation decisions or techniques. Nor is it with the constraints on users' ability to spread information. Nor is it in the robustness of the technical tools that consumers have at their disposal to control their online experiences.

The main problem lies in the rose-colored vision of what companies are doing when consumers log in to social media and other user-facing applications. The claim that these businesses operate platforms for free speech, innovation, and democratic self-government is distracting to the point of making it very difficult for policymakers to see social media for what it is: companies whose ambition is to hold consumers' attention, sometimes at those very consumers' expense. What is worse, the free speech framing has been contorted into a regulatory approach that effectively immunizes companies' bad behavior.

Before spelling out how the doctrine has taken its current shape or offering an alternative approach, I turn in this chapter to how the conventional view, on the right and the left, gets things wrong.

The Unfettered Internet Against the "Woke Mind Virus"

For right-wing activists, the companies' aggressive content moderation policies threaten free speech and the free flow of user-generated content. The first notable time that conservatives raised this point was in the wake of social media bans on high-profile right-wing and MAGA activists during the 2020 presidential election. The distrust intensified when Facebook and Twitter indefinitely suspended Donald Trump for using their services to dispute the results of the 2020 election and promote the attack on the US Capitol in January 2021. Back then, those companies routinely blocked users who posted disinformation about elections or promoted violence. They would not treat President Trump any differently.

This was a flash point for the political right's decades-old skepticism about the political biases of the media. The ostensibly populist complaint about anticonservative contempt goes back at least to Father Coughlin's radio show during the New Deal, Senator Joe McCarthy's attacks on CBS's Edward R. Murrow in the 1950s, President Nixon's criticism of the Washington press corps in the years leading to his impeachment, and 1990s Republican presidential candidate Patrick Buchanan's culture war.

Today, however, it is not just right-wing activists who have said so. Concern about political bias at social media and news organizations generally has grown into a closely held faith among mainstream Republicans. According to a 2017 survey by the Cato Institute, the right-leaning libertarian think tank, 63 percent of Republicans agreed with the view that journalists are, as then-candidate Trump proclaimed, an "enemy of the people."

Prominent free speech groups like the American Civil Liberties Union have also joined this parade of grievance. Their concern has been with "the unchecked power" that social media companies exert over the information that voters receive, particularly when those companies demote or block posts by elected officials and prominent public figures. The companies, they insist, undermine the marketplace of ideas and democratic self-governance. The ACLU, the Electronic Frontier Foundation, and other free speech groups envision a world in which people deliberate over the major issues of the day without hesitation or interference. Platforms, they argue, must allow all perspectives to meet and compete in the marketplace of ideas.

In the US, stories about the way in which the Biden White House bullied or "jawboned" social media during the COVID pandemic to block "misinformation" about health risks fueled this concern. The claim about liberal elites squelching dissent resonated with the running narrative about anticonservative bias—so much so that, in 2023, the Republican attorneys general of Missouri and Louisiana, along with five individual social media users, brought a federal lawsuit in which they alleged that, from 2020 to 2023, President Biden's administration had pressured social media companies to "suppress" conservative viewpoints on election- and COVID-related matters in violation of the First Amendment. They argued that warnings from senior US government officials effectively forced those companies to censor constitutionally protected speech. The Supreme Court eventually rejected the states' arguments because the plaintiffs had failed to allege any actual connection between the government officials' efforts and the social media companies' moderation decisions. After all, many of their takedown and labeling policies preceded President Biden's election.

Civil society groups and human rights advocates abroad have similarly decried what they perceive as governmental orders to deplatform third-party posts that they find concerning. In early 2023, the Indian government demanded that YouTube and other major social media take down a BBC documentary that expressed criticism of President Modi. And, in summer 2024, a prominent Brazilian judge ordered YouTube to block toxic disinformation about elections.

But, in the US, at least, the contemporary grievance about anticonservative bias does not match reality. First, even if you allow that racism, misogyny, and disinformation about public health are views worthy of airing, right-leaning users are not the only ones whom social media has censored. The companies have also banned notable liberal and left-leaning commenters for vitriolic content. Perhaps the most notorious was comedian Kathy Griffin, who was mercilessly scolded for promoting images of a decapitated Donald Trump.

Second, and more important, right-wing activists have substantially overstated their disadvantage. Conservatives actually operated many of the most popular accounts on social media for years. Among the five most popular "news brands" on Instagram over the past few years, for example, are the avowedly conservative Daily Wire (founded by hard-right influencer Ben Shapiro), the *New York Post* (the staunchly conservative daily tabloid owned by NewsCorp), and Newsmax (the far-right cable news and commentary company). These results are notable given that, according to the Pew Research Center, almost a third of people in the United States regularly get their news from social media, including from services like TikTok. Another quarter of people rely on YouTube.

Concerns about anticonservative bias on social media also significantly discount that, over the past several years, a handful of entrepreneurs have developed alternatives to the major social media companies. These include Telegram, Parler, and Trump's own Truth Social. Many of these companies have drawn "alt-right" or far-right users who abide by extremist views and fringe conspiracy theories.

There is probably no better example of this phenomenon than billionaire businessman Elon Musk's acquisition of Twitter in October 2022. Almost immediately after buying it for an eye-popping $44 billion, he renamed the service "X" to bring it in line with his sprawling corporate brand. This was not a shrewd play to make money from microblogging. His investments in PayPal, SpaceX, and Tesla had already made him the richest man in the world. He bought Twitter, he explained, because the company that once touted itself as "the free speech wing of the free speech party" was now stifling conservatives.

Musk held true to his promise as soon as he took over. He fired top executives and dismantled the company's division responsible for combating disinformation and bigotry. He also ordered the immediate reinstatement of the accounts of several high-profile figures on both the right and the left. (The company later again banned comedian Kathy Griffin because she allegedly violated X's rules against impersonation when she mocked Musk.) He also started posting and reposting right-wing policy prescriptions, memes, and conspiracy theories. In the months before the 2024 presidential election, the company seemed to boost his posts to reach users beyond his already massive number of followers. He did not limit himself to the US either. For example, he openly promoted Germany's far-right Alternative for Germany (AfD). In late spring 2025, almost as soon as the president announced a controversial exception to his administration's moratorium on refugees for white South Africans, Grok, X's AI-powered account, crudely amplified the racist worry about "white genocide" in South Africa. Perhaps most ironically, someone at X also prompted the Grok account to "ignore all sources that mention Elon Musk/Donald Trump spread misinformation."

Musk also claimed to see commercial opportunities that he believed the prior leadership had imprudently forgone. Soon after the acquisition, for example, X offered users the chance to pay for "verified" accounts (which get more visibility) as well as more character space to post content. This, he hoped, would help to weed out automated bot accounts and promote authentic free expression. Although it's true that since the change in ownership, X has featured a wider variety of information, the comment threads on the site do not look anything like authentic conversation among users who respectfully disagree. To the contrary, the new verified account strategy has accelerated the proliferation of bot accounts and opened the door to even more lies and bigotry. Inauthentic online content flows freely, as does disinformation about current events, including elections, health and climate science, and war. In the end, Musk's changes have combined to create a perfect environment for AI-powered bot armies to promote Republican candidates and conservative and ultraconservative causes.

Today, with Trump again as president, the right's campaign against anticonservative bias continues as aggressively as ever.

The people he appointed to head the Federal Trade Commission and the Federal Communications Commission repeated and even amped up concerns about anticonservative censorship. They have promised to use their regulatory powers to "fight wokeness" and "smash the censorship cartel" at Big Tech, in spite of major constitutional constraints and their respective agencies' statutory limits.

For example, the current chair of the Federal Trade Commission opened an inquiry into "how technology platforms deny or degrade users' access to services based on the content of their speech or affiliations." The agency cited President Trump's executive order on "Restoring Freedom of Speech and Ending Federal Censorship" as its inspiration. Media Matters, the liberal advocacy organization that is also the subject of an FTC investigation, has successfully blocked the agency's inquiry as a thinly veiled "campaign of retribution" that seeks to chill it and others from criticizing X and other powerful tech companies. Even so, it remains effectively far more muted than it would be given the continued threats and harassment.

The anti-wokeness campaign has paid off. In January 2025, just a couple months after President Trump's election, Meta announced that it would discontinue its partnership with third-party fact-checkers and liberalize its policies against hate speech. Mark Zuckerberg, the company's CEO, explained that Meta's "fact-checkers have just been too politically biased and have destroyed more trust than they've created." The company's new policy would rely more heavily instead on "Community Notes," a tool that crowdsources content moderation by allowing individual users to add caveats and context to contentious posts. (X now already uses this feature.) A user note runs alongside the post at issue if enough other people from different backgrounds rate it as helpful.

Some observers might mistake Meta's recent change as an act of brave defiance against "wokeness." A return to the principles that Zuckerberg set out in his 2018 Georgetown speech. But, in fact, the company has been loosening its policies on disinformation since 2023 by, among other things, serving content and ads about the illegitimacy of the outcome of the 2020 presidential election. The January 2025 announcement about disbanding fact-checking feels more like a performative act of submission; the company has openly chosen to align itself with the country's rightward shift.

The conflation of moderation policy and right-wing politics has been a perfect realization of the aspirations of Silicon Valley libertarians. Peter Thiel, a Silicon Valley investor, is perhaps the most influential among this group. A high-powered Republican megadonor today, he has been a proud conservative since his days in law school in the early 1990s. (In 1995, he co-authored *The Diversity Myth,* a book that attacked political correctness at universities.) He perhaps more than anyone else has aligned Silicon Valley libertarianism with right-wing electoral politics. In *Zero to One*, a book that is both a how-to manual for business start-ups and a libertarian manifesto about free markets, Thiel conveys deep suspicion toward government bureaucracies—for example in education, healthcare, and finance—as well as toward large corporate organizational structures that impede individual creativity.

So, in spite of what we may hear about anticonservative censorship and liberal bias, the major layers of protection for online services continue to prevail in the United States today.

And, contrary to the criticism, conservative voices are as powerful as ever.

Corrupting the Digital Public Sphere

This gets us to the principal concern on the left. Progressive advocates worry that the laissez-faire regulatory approach has allowed hateful people to spread racial and ethnic prejudice, misogyny, and disinformation. This content, advocates on this side argue, may not be illegal, but it should have no place in public life. People of color, young women, immigrants, and others who are members of groups that have been historically marginalized are likelier than others to feel chilled by user-generated content on unmoderated platforms. This content, they argue, does not constitute speech as much as deeply consequential acts of misdirection and subordination. The concept of free speech should not be invoked to punch down or dehumanize people based on race, ethnicity, gender, religion, or other protected categories. Nor should it be used to spread lies or damage civility. Courts, policymakers, and the platforms' content moderators, these advocates argue, should design protections for people who risk the most by "speaking truth to power."

Many of the most successful political campaigns in recent election cycles around the world openly circulate dark rumors and divisive conspiracy theories that tap into inchoate prejudices and social anxieties—sentiments that make people flock to social media. Examples abound. India's Narendra Modi and Romania's Călin Georgescu relied on social media influencers to spread their respective brands of right-wing nationalism and economic populism. Most election reports indicate that Donald Trump's campaign did the same. His campaign reportedly targeted swing voters who consume streaming content. Trump and his running mate's shocking claims about violent immigrants and subway crime sprees in the cities, even if repeatedly disproven, kept voters on high alert. Maybe there is lighthearted mischief in the lies about pet-eating immigrants. But it also resonated with fears about immigration, economic instability, and "replacement."

For decades, writers and scholars have also lamented the ways in which social media corrodes a shared sense of community by slicing and dicing audiences into echo chambers. Some of these writers point longingly to the salons of the European Renaissance. The more common move is for critics to hold up the middle of the twentieth century, when, as the political scientist Robert Putnam recalls, people joined bowling teams and volunteer associations. The broadcast media, on this account, was a source of unity, not polarization.

For this camp, people like CBS News anchor Walter Cronkite conjured a sense of common purpose among viewers across demographic groups, regions, and political party affiliations. On their telling, this was a simpler and better time, when the evening news, as well as Top 40 radio stations and episodic comedy shows like *The Lucille Ball Show* and *The Honeymooners*, supplied Americans with shared cultural idioms. Putnam's *Bowling Alone* is as full-throated a complaint in this tradition as any other.

Unlike the concern about the spread of disinformation, bigotry, and misogyny, this lament feels fanciful and overwrought. After all, voter turnout over the past thirty or so years has been on par with 1960s levels. The argument also presumes that everyone who listened to or watched broadcast programs in the heyday of broadcasting impassively took in the stories the networks told. Communications and cultural studies scholars have been disproving this account for a while now. The Black middle-class protest of the *Amos 'n' Andy* radio program in the 1930s strongly suggests otherwise, for example.

The critique also tends to minimize the extent to which the ad-driven political economy of mass media in the twentieth century incentivized the "Big Three" broadcast networks to be objective. CBS did not cast Cronkite necessarily because he reported the news without bias or opinion. (His sign-off every night—"And that's the way it is"—performed his claim to objectivity.) It and the other networks aspired to objectivity because that narrative form was least likely to alienate most viewers; viewership translated into advertising revenue.

In fact, polarization and partisanship in media are actually almost two centuries old. The penny presses of the nineteenth century were deeply partisan. At the turn of the 1890s, rivals Joseph Pulitzer and William Randolph Hearst built newspaper empires that used sensational headlines and "yellow journalism" about violence, scandal, and corruption. There is nothing new in this phenomenon.

Never mind the trumped-up concerns about censorship. Never mind the illusory faith in genuine democratic debate in the digital public sphere. Demagoguery, alarm, mendacity, and toxicity thrive on the internet today because they hold attention, which translates into big profits for the companies and their partners. The prevailing hands-off regulatory approach to online consumer-facing services has made this so.

If we are to get to the bottom of things, we would be better to understand the incentives that drive the companies to do what they do. This book aims to do this. In the following chapter, we start at the beginning, when the internet was new.

Founding Fathers

Today's hands-off regulatory approach to online consumer services was not inevitable. It took some doing. Ironically, perhaps, the government was essential from the start.

In the early 1960s, at the height of the Cold War, the US Defense Department commissioned researchers at the Massachusetts Institute of Technology to design a national communications system that would survive a catastrophic attack on key mainland military assets and facilities. These researchers responded with a proposal for a geographically dispersed "internetworking" system whose "users" could access the same data at the same time. The system would be "resilient" and "redundant"; this is to say that the information would remain available to all users even if another computer in the network was destroyed. And unlike telephony or cable television, the transmission protocols on which the system relied would obviate the need for a central operator to manage the delivery of data to the intended destination.

Over the next decade or so, the research lab behind the Defense Department's most groundbreaking technological innovations, the Advanced Research Projects Agency, collaborated with researchers at UCLA, UC Santa Barbara, the University of Utah, and Stanford to hone the new technology. It also expanded this new system for "internetworking," now called "ARPANet," to include researchers from a variety of universities and institutions across the US and abroad. The National Science Foundation soon joined and then took over its administration. Under its leadership, and with investments from IBM and MCI, the NSF opened the system, now called the "NSFNet," to an even wider set of research and educational institutions.

By the mid-1980s, online "information services" like CompuServe and Prodigy offered email and other messaging applications, news features, stock quote updates, weather reports, as well as online shopping, chat forums, and games. But it was not until Tim Berners-Lee, an English engineering researcher at the Conseil Européen pour la Recherche Nucléaire (CERN), developed the World Wide Web in 1989 that the internet's potential for widespread adoption became inevitable. Perhaps even more than email, Berners-Lee's user-friendly invention helped the internet to grow beyond a boutique information-sharing network. Among other things, people could go from one site to another simply by clicking a "hyperlink." This mechanism made it easy for newsgroups, chatroom publishers, and early users to promote and connect with each other.

CERN made its web browser available to the public in 1993, but the internet did not really go pop until 1995, when the NSF stepped back from managing the system to allow private companies and private individuals to assume control of the entire internet supply chain—its backbone infrastructure as well as consumer-facing websites and applications like email. In that moment, the public internet that we know today was born. And it was instantly a sensation. It went from having around 7,500 registered domain names in 1993 to two million in 1998, when the NSF finally relinquished the last bit of control to the commercial sector.

A handful of companies stood out in this period, including full-text webhosting services like Altavista and search engines like Lycos. But America Online, which started in the mid-1980s and rebranded in 1989 as a dial-up service, was the standout, outpacing even CompuServe or Prodigy by the end of the nineties. (It actually acquired CompuServe before the end of the decade.) Among other things, it offered email (famously remembered for its "You've Got Mail" alert), as well as the popular AOL Instant Messenger starting in 1997.

Engineering Standards, Cultural Norms

Vinton Cerf was one of the key ARPANet developers as a graduate student at UCLA. In an interview decades later, Cerf recounted that, in the 1960s and 1970s, he and his young colleagues had no idea how transformative networked computing would become. On the one hand, they understood that computers "could be used to augment human intellect." They had learned as much under the tutelage of the acclaimed networked computing pioneer Leonard Kleinrock. On the other hand, he explained, none of them "had any idea what the implications of this work would be in the long run." They did not fully appreciate that focusing on end user control would be as powerful as it has been. The new applications of the

1990s, he marveled, "are blind to nationality, proprietary interests, and hardware platform specifics" largely because he and his young developer cohort sought to ensure that end users could innovate online without permission from a central authority.

End user empowerment, as it turns out, was far more than a purely technical principle. It also resonated with the counterculture of the late 1960s and 1970s. In the Bay Area, end user empowerment was an idea that you were as likely to hear people mention at street fairs as computer science labs. According to Stanford sociologist Fred Turner, the concept helped to "create . . . the conditions under which microcomputers and computer networks could be imagined as tools of liberation" outside of the computer lab.

The *Whole Earth Catalog* embodied this ethos. Based out of Menlo Park, California, in the heart of Silicon Valley, in the late 1960s and early 1970s, it published reviews and articles about products and books. Its legendary founder and publisher, Stewart Brand, sought to index all of the "tools" that would help its "users" to learn "what is worth getting and where and how to do the getting." At first glance, the *Whole Earth Catalog* looked like the far more conventional retail mailers from Sears or L.L.Bean. But on closer inspection, it was very different. It did not advertise or promote products for purchase. It instead offered do-it-yourself life hacks. As Brand explained it, the *Catalog* helped the "user" "conduct his own education, find his own inspiration, shape his own environment." It offered "an alternative view of how to organize and share information," focusing "on individuals rather than the commercial brands."

This ethos would deeply affect the first generation of internet entrepreneurs and engineers. The *Whole Earth Catalog*, according to technology writer John Markoff, was "the internet before the internet. It was the book of the future. It was a web in newsprint." Steve Jobs, the late founder of Apple, explained that the *Catalog* "was sort of like Google in paperback form, thirty-five years before Google came along." It was, he explained, "one of the bibles of my generation." He fondly remembered the publication's mantra: "Stay Hungry. Stay Foolish." That was an urgent call for users to always be adventurous and searching.

Brand transcoded this ethos into the online setting with the Whole Earth 'Lectronic Link (WELL) in 1985, an online community for local Bay Area computing enthusiasts. As the publisher of the *Whole Earth Catalog*, he curated the information and tools about which users learned. With WELL, however, Brand left the content to its users, who eagerly shared their own life hacks, snarky comments, and countercultural worldviews. Soon, early internet users everywhere began to experiment with Usenet and other multi-user chatting protocols to convene similar online groups.

By the early and mid-1990s, a wide range of writers and entrepreneurs helped to spread Brand's gospel of end user empowerment. Microsoft CEO Bill Gates remembered how momentous it all felt when reflecting on his first introduction to online communities like WELL. He and his friends understood that they were "right at the center of the true information revolution." Dreams about "information available at your fingertips and instant global communication" had become a reality. The communications scholar Manuel Castells similarly captured the wonder that people felt at the time in his three-volume collection on the "Information Age." The internet, he wrote in the final volume, had started to transform "the way we think, we produce, we consume, we trade, we manage, we communicate, we live, we die, we make war, and we make love." Users were now in charge. And they were governing online communities with intention and care.

Many of these early adopters believed this emergent user-focused ethos and its related conventions could be exported to people around the world. The internet, communications scholar Lewis A. Friedland wrote in 1996, "connotes a radically new form of democratic practice modified by new information technologies." Justin Hall, an early blogger who later became a popular writer for *Wired*, explained that the internet would make everyone a storyteller. This transformation, he mused, would foster empathy across communities around the world. He and many others believed that the internet's interconnectedness would redound to the benefit of the people who are most underserved.

Similarly, Nicolas Negroponte, the founder of MIT's Media Lab, believed that the internet would enable communities to speak a new "common language" across geopolitical and cultural boundaries. The generations of people who would grow up with the internet, he predicted, would be freed of the "baggage of history," "old prejudices," and "geographic proximity as a sole basis of friendship, collaboration, play, and neighborhood." He predicted that "networking technology and reduced computer costs [would] bring to minorities, poor people, and rural America information and opportunities that have previously been reserved for the elite, urban upper class." In the same way that the printing press had "empowered an emerging middle class," the internet, he claimed, would "expand the same trend on a global scale and encompass citizens of every background and interest." This was a vision that promoted community and, importantly, rejected any role for government.

Cyberspace and the American Dream

For early champions of end user empowerment, the internet would level inequalities, elide geopolitical borders, open access to information, and inspire users to form communities free from the dictates of large companies or governments. They described the new technology in ways that evoked Brand's ideas for the *Whole Earth Catalog*. But they also employed metaphors that suggested something altogether otherworldly.

The "information superhighway" conjured images of travelers speeding over and across geographic boundaries. "Cyberspace" similarly evoked the sense of spatial freedom from earthly constraints. As legal scholars David Johnson and David Post wrote, "cyberspace" was a phenomenon without precedent. The rules that govern it must therefore be distinct from those of "physical, geographically defined territories." Most of the people who enthusiastically used this language recognized that there were unknown dangers ahead—much like the Wild West, the internet could be a "perfect breeding ground for both outlaws and vigilantes." But this risk was presumably part of the internet's charm and promise. In this view, any harm is the incidental cost of creativity, entrepreneurship, discovery, and freedom.

Cyberspace and the American Dream from 1994 remains one of the most cogent articulations of the pioneer mentality. Its four co-authors—Esther Dyson, George Gilder, George Keyworth, and Alvin Toffler—were a who's who of Silicon Valley. Dyson and Gilder were among the early generation of major angel investors. In 1998, Dyson would also become the chair of the newly formed

Internet Corporation for Addresses, Names, and Numbers, the nongovernmental organization responsible for managing the assignment of web addresses. Keyworth, a university-trained physicist, was the CEO of Hewlett-Packard. And Toffler was a celebrated futurist and author whose writing, like Brand, was formative for the baby boom generation.

These were high-tech Brahmins—well-to-do white liberals who were whiggish about free markets, innovation, and civil liberties. For them, the internet and the variety of applications that it spawned (the World Wide Web, email, electronic bulletin boards, chatrooms, and newsgroups) embodied all of these values. The essay's subtitle speaks volumes: "A Magna Carta for the Knowledge Age." People, they wrote, would break away from customary institutions and, instead, create a new world that would be freer and more democratic than anything humanity had ever seen before.

While all of the authors were notable in their own right, Toffler's voice reverberated most. Even today, he is probably most well-known for *Future Shock*, a best-selling work of popular sociology that he published in 1970. "Future shock," he wrote back then, "is a disease of change"—"the process by which the future invades our lives." The book's plan was to help readers "come to terms with the future" and "cope more effectively with both personal and social change in deepening our understanding of how men respond to it." Throughout, Toffler was concerned with privatization and commercialization in urban investment, energy development, agribusiness, "robotology," and pharmaceutical science.

In describing technology in particular, Toffler wrote that "[w]e cannot and must not turn off the switch of technological progress." Instead, "we desperately need a movement for responsible technology." "Technological questions," Toffler continued, "can no longer be answered in technological terms alone. They are political questions"; "we cannot casually delegate responsibility for such decisions to businessmen, scientists, engineers, or administrators who are unaware of the profound consequences for their own actions."

Twenty-five years later, Toffler and the co-authors of *Cyberspace and the American Dream* saw an opportunity for "responsible technology" on the internet. In the new information age, they argued, governments and legacy media companies would be less important. Governments in particular, *Cyberspace and the American Dream* predicted, "will be vastly smaller (perhaps by 50 percent or more) than the current one" given the new opportunities for democratic collaboration online. This new reality, the essay prophesied, "is an inevitable implication of the transition from the centralized power structures of the industrial age to the dispersed, decentralized institutions" that administer the internet.

Such lofty claims seem to draw from the same well that prompted political scientist Francis Fukuyama to predict in 1992 that the collapse of the Soviet Union and the end of the Cold War would create unprecedented opportunities for liberal democracy to spread around the world. This vision would be echoed soon after by Bill Clinton's 1996 State of the Union address when he proclaimed that "the era of big government is over."

A Declaration of the Independence of Cyberspace

John Perry Barlow's *A Declaration of the Independence of Cyberspace* from 1996 is another important artifact of the time. Where the tone of *Cyberspace and the American Dream* bespoke the authors' high-tech Brahmin milieu, the *Declaration* is a barn burner. It proceeds from the start with a bold conceit: "Governments of the Industrial World, you weary giants of flesh and steel . . . have no sovereignty where we gather." The internet, Barlow continues, is the "new home of mind" that will empower its users to transcend repressive status quo conventions and institutions.

> We are creating a world that all may enter without privilege or prejudice accorded by race, economic power, military force, or station of birth.
>
> We are creating a world where anyone, anywhere may express his or her beliefs, no matter how singular, without fear of being coerced into silence or conformity.
>
> Your legal concepts of property, expression, identity, movement, and context do not apply to us. They are all based on matter, and there is no matter here.

The *Declaration* owes its bombast to Barlow's bold ambition: to lambaste members of the US Congress for passing the Communications Decency Act, an anti-porn amendment to the 1934 Communications Act and an important plank of the new Republican majority's "Contract with America." While the majority of its provisions focused on promoting competition and liberalizing government regulation of broadcast licensing and media ownership, the new anti-porn provisions, which revised Section 223 of the Communications Act, criminalized the online distribution of "obscene, lewd, lascivious, filthy, or indecent" material. With this language, Congress aimed to stop the online dissemination of "objectionable" material online.

For Barlow, the amendments to Section 223 were futile and repressive. Congress had in the past forbidden the distribution of sexually explicit content by telephone providers, broadcast stations, and cable operators. But the 1996 amendments to Section 223, Barlow argued, went too far. In any event, he insisted, the internet evades the top-down censorship that legislators contemplated. The internet's Cold War–era developers had built a decentralized communications network that would elide censorship and empower all users to discover new ideas and communities, no matter how unsavory to legislators, in spite of efforts to squash them.

Even as focused on Section 223 as the 844-word *Declaration* was, it also celebrated the internet's promise. In this way, it is also substantively, if not stylistically, aligned with *Cyberspace and the American Dream*. Barlow, despite his tone, was a high-tech Brahmin, too, having been an early acolyte of Brand's ideas and an active member of the WELL online community.

Probably the best indication of Barlow's class and cultural ties to these Silicon Valley elites is the notation that he conspicuously penned at the end of his manifesto: "Davos, Switzerland, February 8, 1996." Davos is where, since the 1970s, the World Economic Forum (WEF) has convened its glittery annual winter meeting of presidents, thought leaders, oligarchs, and power brokers. Barlow was literally breathing the same rarefied air as this crowd when he wrote the *Declaration*. Strident and lofty, it reads as if it was written in exactly that place at that time, when globalization and political liberalism felt ascendant.

In 1996, the Forum's theme was "sustaining globalization." This was a concept that attendees and cosmopolitans around the world embraced but that many labor unions, left-leaning academics and activists, and even some mainstream civil society organizations fiercely resisted. For these critics, the prevailing financial system had entrenched economic inequality and accelerated environmental degradation. "Sustaining globalization," for them, was shorthand for more of the same. The *Declaration*, however, is conspicuously silent about the conflicts that were brewing in protests outside of the WEF in February 1996. Perhaps Barlow meant to channel the anti-globalists' disdain for centralized control by legacy institutions. Perhaps he wrote it to enlist those critics to join the new "home of mind." Even if so, however, the *Declaration* does not even pretend to draw that connection in the text. Instead, Barlow articulated a feeling that internet enthusiasts and Davos attendees shared, in spite of progressive fears about unmitigated globalization: They believed that the internet's constitutive principle of end user empowerment would change the world for the better.

Cyberspace and the American Dream and the *Declaration* owe much to the long line of writers in the Anglo-American tradition who have warned that restrictions on information technology undermine learning and deliberation in democracy. John Milton's *Aeropagitica* from 1644 is a classic in this vein. Milton, the great English poet, wrote it in response to a British parliamentary ordinance that required all print publishers (the transformational information tech startups of the day) to obtain a license from the crown. Lawmakers believed that licensing would prevent the spread of blasphemy, libel, and falsity. But this objective, Milton argued, is misplaced because the people who are most likely to be persuaded by bad information are illiterate; most people, he explained, rely on word of mouth to share information. For him, therefore, the licensing ordinance would interfere with "the privilege and dignity of Learning," the opposite of its drafters' intention.

Like Barlow, writers in this tradition have also argued that the threat of government regulation chills tech entrepreneurship. A classic in this line is Ronald Coase's influential essay on the Federal Communications Commission's regulation of broadcasting in 1959. Coase, a future Nobel laureate, argued that the way in which the agency awards broadcast licenses is prone to inefficiency and corruption, and far less efficient than traditional property rules would be. He pointed to the agency's having notoriously awarded a coveted broadcast license to Lady Bird Johnson, whose only notable qualification was being the wife of then–Senate Majority Leader Lyndon Johnson. Centralized government control like this would always be susceptible to "administrative fiat." Classic liberal common law property rules, on the other hand, would allow entrepreneurs to develop technologies and business models that would account for the real-world costs and benefits of using the electromagnetic spectrum.

And, again, like Barlow and his peers, writers in this tradition also did not think that governments were competent enough to understand new communications technologies. A classic in this vein is political scientist Ithiel de Sola Pool's *Technologies of Freedom*. Writing in 1983, he argued that policymakers and courts had contrived a clumsy regulatory system that favored traditional print publishers but imposed more regulation on new video distribution technologies like cable and satellite television. This approach failed to contemplate the ways in which the new technologies, if left unregulated, could enrich the information environment.

Barlow was channeling these concepts when he penned the *Declaration*. He did not name Milton, Coase, Pool, or any other writers. But he was surely inspired by their ideas. Anyway, in 1996, political liberalism and skepticism about government control of information flows felt especially salient.

Constitutionalizing Cyber-Laissez-Faire

The Rise of Platforms

The high-tech Brahmins and cyber-laissez-faire evangelists of the 1990s aimed high. They weren't just trying to influence policymakers in the short term; they sought to transform the ways in which democracy and markets worked for all time. For this worldview to endure, however, they had to do more than lambaste censorial legislators. They had to weave their messianic claims about the internet into the nation's identity and self-conception.

This ambition was not difficult, given the long tradition of political liberalism in the United States. But the language and framing in *Cyberspace and the American Dream* and the *Declaration* had to work their way into how the legal establishment—judges, legal scholars, Big Law partners, as well as policymakers—talked and thought about constitutional doctrine.

In this regard, these early champions of cyber-laissez-faire and user control were extraordinarily successful. Their advocates at the ACLU and some of the top law firms in the country translated these ideas into legal arguments about what the

First Amendment means in a newly networked world. Their plan worked. In 1997, the Supreme Court effectively constitutionalized cyber-laissez-faire and the myth of user control in *Reno v. ACLU*, striking down Section 223, the anti-porn provisions in the Communications Decency Act that motivated Barlow to write his manifesto. The principles of free speech, tech innovation, and skepticism toward government regulation of information flows articulated in that opinion remain the conceptual building blocks of today's prevailing regulatory approach to the internet.

Writing for the majority in that case, Justice John Paul Stevens explained that the internet had none of the trappings of mid-century mass communications technologies like broadcasting or cable. Audiences for those media really had no choice but to receive the content that the executives at Hollywood studios and television networks produced and distributed.

The internet, on the other hand, features

> audio, video, and still images, as well as interactive, real-time dialogue. Through the use of chat rooms, any person with a phone line can become a town crier with a voice that resonates farther than it could from any soapbox. Through the use of Web pages, mail exploders, and newsgroups, the same individual can become a pamphleteer.

Nor, Justice Stevens continued, is the internet necessarily like print publishing. The web is more like "a vast library including millions of readily available and indexed publications and a sprawling mall offering goods and services" for readers, viewers, and listeners. The Constitution says nothing about consumer rights, but the Court read the First Amendment as broadly affirming that democracies require robust information environments in order to keep the people sovereign. And yet, at the same time, the internet is also a "vast platform" from which publishers could reach "a worldwide audience of millions of readers, viewers, researchers, and buyers." In this light, speakers on the internet, too, have free speech rights. Website publishers. Everyday casual users. Everyone.

The Court struck down Section 223 because the statute was insufficiently attuned to this extraordinary scale and variety. Congress was right to be concerned about children's access to "indecent" and "patently offensive" content. But the manner by which they enacted that "compelling" justification chilled adults' lawful expressive activity. "That burden on adult speech is unacceptable if less restrictive alternatives would be at least as effective in achieving the legitimate purpose that the statute was enacted to serve." Nor, importantly, does the statute differentiate between commercial speech or activities. "Its open-ended prohibitions embrace all nonprofit entities and individuals posting indecent messages or displaying them on their own computers in the presence of minors." The Court, Justice Stevens concluded, had no choice but to strike down the CDA because it is "overbroad," vague, and ineffectually tailored to advance the purposes legislators purported to advance. At least, Congress had failed to incorporate parent-administered content filtering, a far less restrictive way of blocking kids' access to unwanted content.

The Court was surely right that Section 223 infringed on far more protected speech than First Amendment doctrine tolerates. Congress in 1996 should have at least limited its focus to the web, or addressed the law exclusively to commercial activity.

Or it could have drafted provisions that promoted less restrictive alternatives like filtering.

But Justice Stevens's description of the new technology also went further than it needed. His account treated *all* internet applications as virtual "soapboxes" on which "users" could express views or access information. He did not convey any appreciation for the variety in interactivity, reach, or function across different applications. His opinion overlooked important differences—for example, between chatrooms and newsgroups on the one hand and email and web publishing on the other.

These differences are important. Chatrooms and newsgroups are characteristically self-contained communities whose members subscribe to shared discursive norms and conventions. Meanwhile, through email and web publishing, authors and publishers can communicate directly with larger audiences who generally receive the information as distributed. SPAM, for example, is an unavoidable nuisance for most mainstream email users. In the 1990s, before email filters had become sophisticated enough, the burden of detecting SPAM was borne by recipients, not senders. SPAM underscores that email serves a wholly different purpose and operates completely differently than chatrooms or newsgroups, which a person or community moderates according to preexisting guidelines.

The same is mostly true for the web-based applications, where consumers cannot "talk back" unless the website publisher allows it; publishers configure their services to distribute content to audiences as they wish. Remember: Section 223's drafters were most concerned with porn sites that distributed information to whomever visited. Parents could not unilaterally shut those services down. They could only use filters that blocked their kids' access.

For the Court, however, all of these applications—chatrooms, newsgroups, web publishing, and email—were akin to "pamphleteering." They all counted as "cyberspace." This conflation was innocent enough. We see that blanket framing in *Cyberspace and the American Dream* and the *Declaration*, too. Like the authors of both, the justices believed that the internet would promote freedom and democracy, irrespective of application.

That was almost three decades ago. Yet the Court's libertarian approach in *Reno v. ACLU* has mostly endured. One of the more recent opinions in this line involved a North Carolina law that made it a felony for a registered sex offender to use social media when he knew that children used the site. North Carolina convicted the defendant after he had posted an innocuous statement about a positive experience in traffic court on Facebook. The defendant appealed and, in *Packingham v. North Carolina*, the Supreme Court sided with him, holding the statute unconstitutional.

In his opinion for the majority, Justice Anthony Kennedy drew on the language in the Court's *Reno v. ACLU* opinion from two decades before. "[S]ocial media users," he wrote, "employ these websites to engage in a wide array of protected First Amendment activity on topics 'as diverse as human thought.'" "While in the past," he wrote, "there may have been difficulty in identifying the most important places (in a spatial sense) for the exchange of views, today the answer is clear. It is cyberspace—the 'vast democratic forums of the Internet' in general, and social media in particular."

Platform Speech

Some progressive scholars and consumer protection activists have criticized the Court's laissez-faire approach. They argue that it threatens to shield commercial activities that the Constitution should not protect—that these protections grossly distort classic democratic free speech theory in service of corporate power. They liken the current doctrine to *Lochner v. New York*, a discredited 1905 opinion in which the Supreme Court held that the freedom to contract trumps Congress's interest in labor-related protections, like minimum-wage or maximum-hour laws.

Among the more notable contemporary targets of this criticism are Supreme Court cases that invalidated limits on financial contributions to issue advertising during elections, pharmaceutical marketing campaigns, and public access channels offered by cable companies. Some critics do not think it is too late for the Supreme Court to course correct. Or, at least, they argue that courts should resist expanding protections any further into, say, the regulation of cryptocurrency and decentralized finance.

Despite these criticisms, the federal courts have mostly stayed consistently aligned with the laissez-faire approach in *Reno* and *Packingham*, extending it to protect a wide range of internet companies. Over the past three or so decades, for example, judges across the country have cited the First Amendment to protect search engine rankings and video-sharing feeds. Free speech under the Constitution, the courts have said, is not limited to the words or images that individuals post or distribute. Online services are internet users, too; they have the right to decide how and which ideas to amplify, demote, demonetize, or altogether block.

This is to say that, since *Reno v. ACLU,* the courts have said that internet companies are constitutionally relevant speech platforms—that, by featuring and moderating user-generated posts and videos, for example, they promote democratic free speech values. In this view, they are not like television broadcasters or cable operators, which mostly control the information that consumers receive. Congress and regulators could require those companies to carry certain kinds of content given their powerful gatekeeping role in local markets for video content. The courts have treated online services differently on the theory that, as they are designed, anyone can be a pamphleteer or publisher.

Almost three decades since *Reno,* we can say that information libertarianism has prevailed. Today, for fear of hampering free speech, courts refrain from making distinctions between beneficial and harmful content. Congress, they have explained, does not want judges to police the manner in which a company distributes or publishes lawful content, no matter how crass, snarky, or unelevated it may seem, because even frivolous exchanges and posts may be generative in ways that are not obvious. The best regulatory approach, according to the prevailing doctrine, is forbearance.

"Censorship-Industrial Complex"

These constitutional protections have not stopped conservative political activists from attacking the biggest platforms for their liberal bias. These companies hold too much sway over public discourse, conservatives argue. They comprise today's "Censorship-Industrial Complex." Policymakers should require companies to carry user-generated content they disagree with, just as they once required broadcasters and cable companies to do.

Some of the most ambitious versions of this attack have come from Republican-led legislatures in Texas and Florida. In 2021, those states passed laws that prohibit social media companies from discriminating against consumers based on viewpoint—really, against what those legislatures perceive as anticonservative bias. The attorneys general in those states argue that the companies' moderation decisions skew the marketplace of ideas in favor of "the West Coast oligarchs." The new state laws would ensure that the largest social media companies do not unilaterally control what people see online.

The two statutes would achieve these objectives in slightly different ways. Florida's law prohibits "social media platforms" that have over $100 million in annual gross revenue or more than one hundred million monthly visitors from "censoring" user posts based on content or source. It specifically forbids censorship of large "journalist enterprises" online, on cable television, and in broadcasting. The Texas statute similarly forbids social media companies with over fifty million monthly active users from interfering with the "information, comments, messages, or images" that their consumers post. Both laws also required companies to explain each moderation decision that "censors" or discriminates against consumers based on viewpoint.

NetChoice and the Computer & Communications Industry Association, the industry groups that represent the largest and most powerful internet companies in the world, immediately brought constitutional challenges to both state laws. They argued that the statutes would require companies to distribute speech with which those companies did not agree. Such state-imposed must-carry requirements, they asserted, intruded on the companies' constitutional right to speak and curate content as they see fit.

The industry groups' challenge was as aggressive as they come. In their suit, they argued that few, if any, potential applications of the state provisions were lawful. They also challenged the statutes' disclosure and transparency requirements as unduly burdensome. Companies, they argued, would be forced into refraining from moderating user posts if, as the laws require, they must explain each of their moderation decisions. At the moment, because of different decisions by different federal circuit courts, the Texas law stands and the Florida law has been struck down.

When railing against social media liberal bias, many contemporary conservatives have approvingly referred to two older cases. In the first, *Pruneyard Shopping Center v. Robins* from 1980, security guards in a small city outside of San Jose, California, removed a student-led petition-signing campaign from a shopping center because the students had not gotten prior approval from the business's management. The California Supreme Court determined that the state constitution's free speech provision required shopping center owners to allow expressive activity on their property. In 1980, then-Justice William H. Rehnquist penned the US Supreme Court's unanimous opinion in *Pruneyard,* siding with the students. States, he wrote, may adopt constitutional rights that are "more expansive than those conferred by the Federal Constitution." In the second, *Rumsfeld v. Forum for Academic & Institutional Rights,* law schools brought a First Amendment challenge to a federal law that conditioned federal funding on allowing military recruiters the same access to students as nonmilitary recruiters. FAIR, on behalf of some universities, did not want to allow or associate themselves with such outreach efforts due to their disagreement with the military's policy on homosexuals in their ranks, colloquially referred to as "don't ask, don't tell." The Court rejected the universities' challenge, holding, as in *Pruneyard*, that the statute at issue did not limit what schools may say about the military's employment policies.

Justice Clarence Thomas cited *Pruneyard* in a 2019 opinion to speculate that "some digital platforms are sufficiently akin to common carriers or places of accommodation to be regulated in this manner." President Trump's White House relied on *Pruneyard* in its 2020 "Executive Order on Preventing Online Censorship" after Twitter had labeled two of his tweets "potentially misleading." Former Republican presidential candidate Vivek Ramaswamy co-authored a 2021 opinion piece in the *Wall Street Journal* in which, channeling the *Pruneyard* reasoning, he argued that "Google, Facebook and Twitter should be treated as state actors under existing legal doctrines." And Eugene Volokh, the libertarian First Amendment scholar, has written that some private online platform "functions" may be sufficiently imbued with a public character to justify free speech rights for individuals who use those functions.

Hopeful that they would find a receptive majority, conservative leaders in Texas and Florida appealed to the Supreme Court, hoping for a strong affirmation of the principle that content moderation violates the First Amendment. Of course, NetChoice and CCIA also appealed, hoping for the opposite result. On the last day of the 2023–24 term, Justice Elena Kagan, writing for the Court, effectively agreed to reverse Texas's ban on content moderation and to permit Florida to keep moderating content. But, she explained, the Court was returning both cases to the lower courts for reconsideration because they had not sufficiently considered the full range of online functions and features that might be implicated by the laws.

So now the *Moody v. NetChoice* case has gone back to the lower courts in Florida and Texas where the judges there must determine whether there are enough unconstitutional applications of those states' laws to render them invalid. In many regards, this is a refreshing turn to a more careful and iterative form of policymaking—a jurisprudential posture that hews far more closely to the way in which the Court has analyzed media technologies in the past. It has taken thirty years since *Reno v. ACLU*, but the Court may finally be at an important inflection point. Alas, it may be that the libertarian project under the First Amendment has reached its limit. Decisions by the Court in the first half of 2025 in *TikTok v. Garland* (on a federal ban on certain foreign adversary–controlled applications) and *Free Speech Coalition v. Paxton* (on a state law that requires porn sites to verify their users' age) suggest as much. We will return to those later.

Even given this potential turn to a more nuanced application of the First Amendment to internet applications, online services have been able to avoid legal accountability pursuant to another powerful legal authority. This one, however, is a statutory liability shield, not a constitutional protection. And, as with First Amendment doctrine since *ACLU v. Reno*, courts have been essential to defining the liability shield's extremely generous scope for just as long. We turn to that next.

Codifying the Engagement Model

The Liability Shield

Companies and their boosters have been effective at persuading courts that consumer-facing services are constitutionally protected platforms for free speech. But that is not the only tool that they have at their disposal. They also enjoy an exceptional shield from civil liability for distributing user-generated content. Its advocates summon the same cyber-laissez-faire framing that now defines First Amendment doctrine.

Indeed, even before the Court announced its opinion in *Reno v. ACLU,* Silicon Valley advocates in the 1990s had called on Congress to enact a statutory protection that would effectively immunize them from liability for "publishing" the unlawful content of their third-party users—a broad protection that, with the possible exception of gun manufacturers, no other industry enjoys. Only then, they believed, could the internet truly realize its full potential. The First Amendment was not enough.

Legislators in the House and Senate could be forgiven for feeling whiplash. On the one hand, the anti-porn advocates wanted policymakers to limit the kinds of information that appeared online. That is what Section 223 of the Communications Decency Act, which the Supreme Court struck down in 1997, would have provided. On the other hand, Silicon Valley advocates wanted companies to have the freedom to distribute third-party content, including lawful porn, without limitation. This is what the new provision they sought would do.

As irreconcilable as the two provisions may seem, both ambitions prevailed in the Communications Decency Act. Alongside Section 223, the law also included a liability shield for online companies that distribute third-party content. This provision is called Section 230. Through this provision, Congress sought to ensure that a "vibrant and competitive free market" would flourish, "unfettered by Federal or State regulation."

The statute's main provision—Section 230(c)—contains three operative protections. The first—Section 230(c)(1)—is short: "No provider or user of an interactive computer service shall be treated as the publisher or speaker of any information provided by another information content provider." Under this subsection, a court may only hold such a company liable if that company is not in any way responsible for creating or developing third-party content. More than the statute's two other protections, Section 230(c)(1), as we will see, has had a decisive impact on the internet as we know it today.

The other two subsections, both under Section 230(c)(2), have been overshadowed significantly by (c)(1). Section 230(c)(2)(A), the first of the lesser two, establishes a "Good Samaritan" protection for online services. The title comes from the familiar biblical parable about a do-gooder who helps a gravely injured man on the side of a road. The relevant scriptural account tells the story of two passersby, a priest and a Levite, who, one after the other, fail to help a man who has been robbed, beaten, and left for dead on the side of a road. The two fail to help because of prevailing religious teachings against defiling a corpse and avoiding uncleanliness. A third passerby from Samaria tends to the man's wounds, takes him to a nearby inn, and pays the innkeeper for every day that the victim stays to recover.

States across the country have enacted laws that define the duties that people owe to strangers who are in distress. These Good Samaritan laws mean to create an incentive to do good rather than impose the duty to do so. A defendant will invoke such a statute when a plaintiff who was in some mortal danger alleges that their injuries worsened when the defendant tried to help. In cases against emergency medical technicians or physicians, for example, some states require plaintiffs to show, among other things, that the medical professionals were grossly negligent (that is, well beyond merely negligent) when they came to help. Many states extend that protection to anyone who renders any kind of assistance, even outside of an emergency setting.

Congress imported these ideas into the Communications Decency Act, starting with the operative section's header: "Protection for 'Good Samaritan' blocking and screening of offensive material." That provision, section(c)(2)(A), shields "interactive computer services" that voluntarily block or take down "objectionable" third-party content. Congress assumed that, freed from the threat of litigation, online companies would go out of their way to employ content screening techniques and policies. Section 230(c)(2)(B), meanwhile, establishes a protection for companies that make technical tools, including content filters, available to parents who want to limit their kids' access to objectionable material. "Parents and families," Congress reasoned, "are better suited to guard the portals of cyberspace and protect our children than our Government bureaucrats." They should not be punished for making filtering tools available to their users.

Silicon Valley advocates were never concerned about the harms of porn or "objectionable" content. To the contrary, they were worried that the courts would punish internet companies that failed to block content from third parties. The companies' worry was not fanciful. In 1995, a trial court in Long Island determined that Prodigy, an early online service, was liable for defamatory content that its users posted. Silicon Valley advocates believed that the internet could not thrive if online service providers like Prodigy could be sued for any single third-party post among the massive amounts of content that they distributed. Four years earlier, a federal trial court in New York had ruled that a similar service, CompuServe, could be held liable for defamatory third-party content that it knowingly distributed.

The Prodigy and CompuServe cases dismayed Silicon Valley advocates. They looked to Capitol Hill for help. There, they found two champions in the US House of Representatives—California Republican Christopher Cox and Oregon Democrat Ron Wyden—who proposed a last-minute amendment that would effectively overrule the decisions in *Prodigy* and *CompuServe*. While it would not ignore the threat that certain kinds of content posed, the new law, Representative Cox explained, ensured that "a Federal Computer Commission with an army of bureaucrats" would keep their hands off the internet. The new technology, he insisted, had "grown up to be what it is without" government interference. (His remarks said nothing about the indispensable roles of the

Departments of Defense and Commerce in the decades before.) The liability shield, he predicted, would "encourage what is right now the most energetic technological revolution that any of us has ever witnessed."

Cyberlaw: Tech Rules in the Shadow of Law

Representative Cox may as well have cited Barlow, Toffler, and the other early champions of cyberspace. Indeed, the pioneer generation's messianic ideology influenced policymakers in a handful of other new statutes as well, of which the liability shield was really just the first.

The Internet Tax Freedom Act of 1998, for example, barred state and local taxes on electronic commerce for three years. Congress reasoned that, without this protection, a patchwork of state tax laws would impede internet commerce, which is necessarily interstate. The Digital Millennium Act of 1998, meanwhile, lightened traditional publisher obligations under copyright law for online services. Through it, Congress required "online services" to block or take down copyrighted work that individual users post without permission, but the burden of monitoring third-party posts of unauthorized content would fall on copyright owners, not online services. The passage of these laws underscored that, for Congress, the internet was different enough to justify a set of rules that were dramatically more forgiving than those for print, broadcasting, cable, and other twentieth century media technologies.

Congress, however, was not always *completely* deferential to online companies during this early period. They repeatedly drew a line at children. Again, remember Section 223, which the Supreme Court struck down in 1997 just a year after Congress had passed it. That failure did not stop Congress from continuing to try to protect children from online porn. The Supreme Court struck these laws down as unconstitutional each time.

The one online child protection law to survive judicial scrutiny at this time was the Children's Online Privacy Protection Act of 1998. To this day, it imposes restrictions on the ways in which companies direct online content to children and collect personal data.

Even this law, however, set out a relatively hands-off approach. In many ways, it is the exception that proves the rule. The statute does not impose time-of-day restrictions on the distribution of objectionable content in the same way, for example, the Communications Act does for broadcasting or cable. It instead allows parents to control their children's access to online services; kids could gain access to these services if their parents allowed it. Users would be the ones to decide.

This last point underscores Congress's frame of mind in the late 1990s. Together with Section 230, the other laws reflected a deferential model for legislating the internet. These "cyberlaws" did not so much impose firm legal duties on online companies as much as establish safe harbors and immunities that helped internet companies *avoid liability* as long as they ostensibly empowered consumers to make choices about their online experiences.

A handful of supportive legal scholars added intellectual heft to this approach. Drawing on the themes in *Cyberspace and the American Dream* and the *Declaration of the Independence of Cyberspace*, these "cyberprofs" posited that the internet is a democracy-enhancing communications infrastructure that elides conventional governmental regulation. These scholars argued that, in order to be effective, government regulation would have to be more hands-off than twentieth-century communications law and policy had been. For them, command-and-control regulation of broadcasting and cable television were not the right fit.

David R. Johnson and Robert Post were leaders of this vanguard of legal scholars. Their seminal law review article, "Law and Borders: The Rise of Law in Cyberspace," drew from the same ideological well on which Barlow relied for his *Declaration*. The internet, the two argued, transcends "territorial borders, creating a new realm of human activity and undermining the feasibility—and legitimacy—of laws based on geographic boundaries." This new "space" would "create its own law and legal institutions." Policymakers, Johnson and Post assured their readers, should not feel threatened by these new developments. To the contrary, "territorially based lawmakers and law enforcers" should "learn to defer to the self-regulatory" modes of governance already at work online. "These new rules," they argued, "will play the role of law by defining legal personhood and property, resolving disputes, and crystallizing a collective conversation about online participants' core values."

Some people were not so swept away. These skeptics did not think the rules of the road for the internet should be any different than those for other industries. In a widely cited speech at a University of Chicago conference on the "Law of Cyberspace," federal appeals court Judge Frank Easterbrook admonished attendees as engaging in a kind of self-important "multidisciplinary dilettantism." "Cyberlaw," he insisted, is as coherent a body of law as, say, the "Law of the Horse." Which is to say, not at all.

Judge Easterbrook's speech immediately triggered a response from cyberprofs. The new technology, they argued, showed that sometimes "system design choices" are better at governing behavior than conventional law. This is why for them lawmakers should turn away from heavy-handed prohibitions and instead adopt flexible default rules that promote user choice. That kind of approach would do better than hard-and-fast prohibitions at achieving public policy norms like privacy or copyright protection.

Legal scholars have for decades considered the ways in which social ordering emerges in the "shadow of the law," rather than directly pursuant to it. In this framing, law is not the only source of norms, and formal legal process is not the sole forum for dispute resolution. Rather, people often see legal rules "as a fact to be taken into account rather than as a normative framework" that consumers must follow. People articulate and enforce norms at home, in their communities, in the workplace, and other settings through what Marc Galanter has called "indigenous ordering."

For example, ranchers in Northern California sometimes settle disputes regarding cattle trespass and boundary fences through informal norm-based rules and sanctions, outside of the legal regime. They would only turn to law as such when disputes arose over highway collisions involving livestock. Similarly, in the completely different setting of divorce proceedings, discretionary legal standards, as opposed to clear rules, "can substantially affect the relative bargaining strength" of the parties due to differences in the parties' attitudes toward risk and ability to bear transaction costs.

Scholars have also shown that law has an expressive function, as opposed to simply a behavioral or penal one. Through it, lawmakers "express a judgment" that means to change norms involving costly behaviors like "smoking, using drugs, or engaging in unsafe sex." Similarly, "choice architectures" are more effective than bright-line prohibitions or mandates. Indeed, sometimes, regulators' opposition to a social norm might undermine enforcement. Accordingly, policymakers could enact measures that "nudge" (rather than direct) consumers and companies to behave in certain welfare-maximizing ways. In transportation policy, for example, speed bumps and other "traffic calming" designs often control car traffic better than speed limits do. In public health, warning labels about sodium or calories at fast-food establishments are easier to police and tend to promote healthier choices than requirements on restaurant kitchens. In education, hallway and classroom design are sometimes better at managing the ways in which students learn and interact with each other than hall monitors.

Legal scholars Lawrence Lessig and Joel Reidenberg drew on these ideas to dispute Easterbrook's critique. Sometimes, they argued, network engineers, computer scientists, and software developers are better at achieving public policy priorities than directives in legislation or regulation are. Website developers, for example, might create innovative browser plug-ins that give users the ability to protect against privacy or security breaches in ways that intellectual property law at the time made prohibited. Music studios could implement "digital rights management" technologies to limit the ways in which people copy and share copyrighted works in ways that traditional legal processes and government regulation made difficult.

Traditional regulatory measures would continue to be useful, Lessig and Reidenberg acknowledged. But policymakers should draft them in ways that encourage engineers, scientists, and developers at companies to come up with technologies that advance important public policy objectives. Governments, for example, could impose liability on various network actors for failing to give users the option of sharing personal information. Or they could create flexible safe harbors for specific practices, like the Digital Millenium Copyright Act's notice-and-takedown regime for copyright violations. Or policymakers could sanction users who circumvented certain technical measures that protect intellectual property. The key was to give developers the incentive to design technologies that advanced policy priorities like user control and intellectual property protection.

Section 230's liability shield, as well as the safe harbors under laws passed during the same period, reflect this deferential "cyber" thinking. Those statutes give companies and users the freedom to choose their own way. Through the liability shield in particular, Congress deferred content moderation questions to services that distributed user-generated content on the theory that those platforms were best situated to know what consumers want. Companies and developers would now have the flexibility to attend to consumer demand without the threat of liability. Online services, Congress assumed, could provide services that accommodated consumers' interest in (or distaste for) all kinds of content in ways that are far more effective than blunt directives or prohibitions could. At least, users could exercise a choice about which platforms or content they wanted to engage. This regulatory framing proceeded on the faith that society was better off when tech developers could innovate freely, no matter the apparent social costs, as long as they supported user-generated content and user choice. Legal rules would just have to give them that discretion and flexibility.

Government Hands Off, Platforms Take Over

As momentous as the other measures were, Section 230 was at the heart of Congress's new cyber-laissez-faire effort. The liability shield was like anabolic steroids for a new generation of internet companies. It immediately spurred developers to design "Web 2.0" applications for social networking, video sharing, and massive news and gossip discussion sites.

Section 230's impact, however, would only become clear after a federal appeals court evaluated the law's application to the Big Tech company of the day, America Online, in November 1997. In that case, an unknown AOL user had posted advertisements for merchandise that smugly celebrated the 1995 Oklahoma City bombing. But there was a cruel twist: The anonymous user attributed the post to Kenneth Zeran, who had nothing to do with the promotion. Zeran predictably received hateful calls and death threats at his home as soon as the fake ads went up. He promptly asked AOL to take the advertisements down and the company complied. The anonymous user, however, continued to post new fake ads over the next few days. Zeran contacted AOL each time and the company abided by his takedown requests every time. AOL, however, refused to issue a retraction or promise to block any future posts given the sheer amount of third-party content on its site. Unfortunately for Zeran, it would be a losing game of whack-a-mole.

Zeran sued AOL in the Northern District of Virginia, where AOL's headquarters sat. He alleged that the company had a duty to take down all defamatory posts as soon as it had notice of them. He also asked the court to have AOL block all future similar posts. The trial court dismissed Zeran's claims, citing Section 230. A panel of federal judges in Richmond, Virginia, agreed.

Chief Judge Harvie Wilkinson, writing for a three-judge panel, determined that Section 230 was a very broad protection. The disturbing facts in the case did nothing to persuade the court otherwise. The company, the panel held, did not owe any obligation to Zeran, even after it had learned about the defamatory content. The purpose of the statute was to promote the "diversity of political discourse, unique opportunities for cultural development, and myriad avenues for intellectual activity." AOL could never play that important role if it had to monitor "millions" of subscribers' online activities. The panel understood that Zeran's life had been turned upside down and that its decision would have far-reaching implications for similar victims. But, it explained, Congress had made its choice.

The court relied heavily on the statute's high-minded prefatory language. Congress, it wrote, wanted the "burgeoning Internet medium" to thrive, undeterred by the chilling threat of litigation. A broad reading of immunity would best achieve these statutory purposes. With this protection, companies could meet consumer demand where they want to meet it—not where legislators or courts required them to. They could be as permissive or restrictive as they wanted, consistent with their own interests and priorities. The law accordingly goes further than the First Amendment to promote platforms for third-party content.

The court published its decision in *Zeran* in the same year that the Supreme Court decided *Reno v. ACLU*, when the public internet was only a couple years old. But the emergent Section 230 doctrine, and Judge Wilkinson's opinion, firmly set in motion a cyber-laissez-faire regulatory framework that would inspire a new generation of consumer-facing services for user-generated content.

America Online, the biggest of the early tech companies at the turn of the century, was the clear beneficiary. As prominent as it was in the late 1990s, it was the target of dozens of major lawsuits, including, for example, cases involving slanderous remarks about a Clinton White House official, an incorrect report on a company's stock prices, and an AOL chatroom through which one user sent a debilitating virus to another. The courts cited Section 230 to reject all of the plaintiffs' claims in these cases because, under the scheme set up by Congress, AOL was a mere platform for third-party content.

But AOL was not the only one to benefit. As antiquated as the electronic bulletin board in *Zeran* may seem to us today, Judge Wilkinson's opinion has held extraordinary influence over how courts across the country apply Section 230 to consumer-facing services in the years since. They cited it to protect search engines, dating and matchmaking services, consumer complaint and advocacy sites, and sites that solicited sordid gossip. Online services have also successfully invoked the protection against claims that go beyond defamation or other reputational injuries, including cases involving false advertising, business-related torts, and breach of contract. In most of these circumstances, the courts have stayed true to the *Zeran* court's expansive reading of Section 230. Along the way, they have given license to an extraordinary number of online services, including home sharing apps, content streaming, multiplayer online games, and crowd-sourced review sites.

This is not to say that defendant companies always win. While the liability shield remains robust, the courts have rejected the Section 230 defense when the service at issue is not simply "publishing" third-party content, but rather engaging in some separate harmful conduct or activity. They have held, for example, that an unfulfilled promise from a Yahoo phone representative to take down defamatory posts created a duty to do so that was unrelated to publishing. Model Mayhem, an online marketplace for aspiring models, was not immune from liability for failing to warn users about two men who used the service to lure and then sexually assault women in hotel rooms. The duty to warn there was unrelated to publishing. Homesharing sites may be liable for failing to verify that hosts have registered with local municipal ordinances as lessors of their short-term rental units. That duty was unrelated to publishing the listings. A marketing network that placed clients' advertisements on affiliated "fake news" sites was not immune for publishing the deceptive product information in advertisements. Google was not immune from consumer protection claims for removing a video in violation of its own terms of service.

"Moral Hazard on Stilts"

Even in light of these important doctrinal refinements, Section 230 doctrine has had an extraordinary impact. The courts have shielded platforms in the vast majority of cases in which plaintiffs sue for harms arising from the service's distribution of user-generated content.

Courts rarely if ever refer to the biblical parable that ostensibly inspired Congress to enact Section 230. How could they? Many platform defendants are unsympathetic to the underlying lesson about doing good. To the contrary, as legal scholar Mary Anne Franks has written, the liability shield promotes "moral hazard on stilts." It has effectively encouraged services "to be increasingly reckless with regard to abusive and unlawful content on their platforms." Indeed, consistent with the themes in this book, she argues, they have invoked Section 230 and the First Amendment to avoid legal accountability.

The current state of the law has left judges befuddled. Judge Easterbrook, who derided "cyberlaw" in the 1990s as nothing more than "multidisciplinary dilettantism," has had to apply the law. But he, as with other judges, remains concerned that the statute has not accomplished what the Good Samaritan title promised. Today, companies do not really seem to care about any special advantage for blocking harmful expressive activity. To the contrary, Section 230(c)(1) has created an incentive to distribute content, whether the "platforms" moderate or not. Its "principal effect," one prominent federal appellate judge explained, "is to induce ISPs to do nothing about the distribution of indecent and offensive materials via their services." The courts have cited the liability shield to protect websites like The Dirty that solicit humiliating gossip and media about young women. For The Dirty, this was actually part of an extortionate business model in which they would require victims to pay a fee to take the offending content down. The courts have also relied on Section 230 to block a case against Backpage, the now-defunct classified site that knowingly featured posts about sex trafficking of minors.

Advocates of regulatory forbearance explain that this is what freedom looks like. A vibrant information environment of authentic user-generated content is good for democratic deliberation. At worst, these advocates assert, occasional consumer harms are the incidental costs of free speech in democracy. In this view, the liability shield allows all manner of content to get an airing, and this is a good thing. Better to allow as much information to flow than the alternative. As one prominent tech and law scholar put it, "The mistakes caused by liability are worse than the mistakes caused by immunity."

There is a glaring irony in the way in which Section 230 has played out since Congress enacted it. Just as the liability shield has given birth to a variety of services, it also has introduced harms that its early proponents surely did not anticipate. I offer here three examples: Match, Armslist, and the Experience Project.

Push Notifications: Match

Match is a popular online dating app. Its subscribers can create profiles and check out prospects. Users can also create a nonsubscriber profile, but they would not have access to services like direct messaging. According to a recent case filed by the Federal Trade Commission, most consumers (subscribers and nonsubscribers) do know that fraudsters operate over a quarter of the accounts on Match. These scammers use the service to dupe nonsubscribers into purchasing subscriptions and then con them into sharing personal information.

In its complaint, the FTC alleged that Match understood this phenomenon, but nevertheless did nothing to stop the scammers from inducing nonsubscribers into purchasing subscriptions. To the contrary, Match emailed notifications and recommendations whenever a subscriber sent an in-app direct message, mindful that a large percentage of those communications originated from scammers. The agency also alleged that the company further facilitated these scams by guaranteeing unwitting new subscribers a "match" within six months. (Consumers would get six free months of the service if they did not find a match.) The FTC's complaint said that Match facilitated scams or at least deceptively exposed consumers to scams, costing them an estimated $844 million in losses.

Match moved to dismiss the complaint, citing the liability shield. The trial judge agreed. The FTC, the court explained, was seeking to hold Match accountable for content that originated from third parties. The company did not create or develop the content. The email notifications are "content-neutral tools" that "automatically" facilitate communications between users. They "are not content in and of themselves."

Black Markets: Armslist

In many regards, the Match case is a relatively straightforward application of Section 230 doctrine. More notorious examples involve ostensibly neutral online marketplaces where consumers meet strangers online to procure dangerous products. This is how morally bankrupt the laissez-faire approach under Section 230 has been.

Armslist is a large "online firearms marketplace." Any visitor to the site can review information and advertisements for guns and related gear. Account holders may post advertisements or "want to buy" requests for specific kinds of weapons. For these people, the company makes no distinction between federally licensed dealers, who must be subject to background checks, and individual private sellers. Armslist facilitates the connection between sellers and buyers through a search function on the site. It also relies on self-regulatory moderation techniques, including user tags for scams, misclassification, and overly high prices. It does not, however, provide a mechanism for consumers to flag illegal activity. It also does not limit who may create an account or post advertisements, which "allows buyers to avoid state-mandated waiting periods and other requirements."

On the site, the company makes a contact tool available through which potential sellers and buyers may communicate with each other directly. It makes money through third-party advertisements but does not earn a brokerage fee or anything similar for the completed transaction.

Armslist does not protect against harm from the third-party information that it distributes. To be clear, such information is not like a social media post about Alicia Keys. It specifically facilitates commercial transactions involving products that could have devastating consequences. As a categorical matter, Armslist traffics in information about how to sell or buy weaponry. It and other sites like it are the destination for people who want to sell or buy unregistered guns.

But under Section 230, it never has to assume the risk for the harmful acts that predictably follow from its account holders' advertisements or "want to buy" posts. That is what happened in Zina Daniel Haughton's case. In 2012, she obtained a restraining order against her husband after he threatened to kill her. Among other things, the order forbade him from possessing a firearm for four years. Her husband nevertheless went to Armslist to purchase "a handgun with a high-capacity magazine 'asap,'" clearly with the intention to make good on his threat to his wife. On the next day, he fatally shot Zina, two of her co-workers, and himself in a suburban Milwaukee hair salon. Zina's daughter, who was a witness to the murders, brought the case against Armslist, alleging that the company was partially, if not wholly, responsible for the murders.

The Wisconsin Supreme Court dismissed the case against Armslist on Section 230 grounds. Armslist, of course, did not know that Zina's husband would use the site as part of his vicious plan. But it did not have to take steps to learn about any specific buyer's intentions, let alone those of the husband. The company, the court held, did not create the content at issue for the purposes of Section 230, even if it understood that certain people could purchase a firearm despite state law restrictions on those buyers. That was not a risk it had to bear. The website did not have to implement controls on who could buy weapons. Nor did it have to employ any mechanisms that would allow visitors or account holders to flag potentially dangerous people on the site.

Free-for-All Chatrooms: The Experience Project

Now take the Experience Project. Unlike Armslist, it was not focused on a specific range of products or activities. It instead aimed to create communities out of the information that anonymous users volunteered. Given the variety of subject matter, its matching technique was far more involved. It relied on algorithms to identify like-minded people. Typing something as simple as "I like basset hounds" or "UFOs are real" would help the Experience Project make the connection. The service would send its users an email notification when someone responded to related inquiries. It made money through advertisements, as well as through donations and the sale of tokens that users could spend to communicate with others in their groups.

As the Experience Project's user base grew, so, too, did the variety and number of its constituent communities. The company, however, had to shut down in 2016 because, according to its founders, "bad apples" were flocking to the site. The more notorious cases included a sexual predator who used the site to entrap underage victims as well as a man who killed a woman he met through the site.

The one episode that probably caused its undoing, however, was the drug overdose death of teenager Wesley Greer. The teen used the site to meet people and find heroin. He died of fentanyl poisoning after unknowingly purchasing heroin laced with fentanyl from another user who posted ads on the site labeled "I Love Heroin" and "Heroin in Orlando." Greer's mother sued the Experience Project for wrongful death. She argued that he would not have obtained the drugs that killed him without its services. But the courts rejected her claim on Section 230 grounds. The Ninth Circuit held that the Experience Project could not be held accountable for Wesley's death because the service merely connected users to people that they sought out.

Self-Regulating for Self-Gain

Section 230's early advocates did not have *these* applications and their real-world impacts in mind. They were not thinking about fraudulent push notifications for dating services, black markets through which people could traffic in unregistered automatic weapons, or chatrooms through which young people could buy heroin. But the law's current defenders have not backed down. They continue to trot out the free speech theories that gave rise to the liability shield in the 1990s. Presumably, according to its staunchest advocates, the tragedies that befell the Daniel and Greer families in the Armslist and Experience Project cases are the unavoidable incidents of free speech and innovation.

The companies know better. Since 2022, in the European Union, platforms must attend to the systemic risks that certain content will harm users because the Digital Services Act requires it. In the US, platforms moderate risk because of pushback from consumers—that is, as a function of classic supply and demand. They are especially aware of how upset users and advertisers get when children get hurt. This is why, for over a decade now, the largest companies have partnered with the National Center for Missing and Exploited Children, to identify and take down child sexual abuse material. They also have recently come together to pledge to eliminate AI-generated material of this kind. Especially alarmed by the heinous 2019 massacre in a mosque in Christchurch, New Zealand, they also launched the Global Internet Forum to Counter Terrorism to collaborate on techniques for flagging and taking down violent or terrorist-related content.

These are not the only ways in which US companies have slowed or stopped the algorithmic distribution and amplification of lethal content. Popular services have introduced friction into the ways in which their consumers share and engage the most dangerous content. "Circuit breakers" like user and automated flagging draw inspiration from the basic idea made popular by Renee DiResta in 2018 that free speech on platforms "is not the same as free reach," particularly for content that is corrosive or otherwise harmful. The companies have accordingly integrated warning labels and other "frictive designs" alongside misleading or dangerous content. Some interventions like these are more effective than others. But research has shown that such prompts dampen users' impulse to share harmful content.

Platforms have also developed institutional processes that convey their seriousness about attending to the balance between promoting free speech and blocking harmful content. Perhaps the most notable is Meta's Oversight Board, which launched in 2020. It adjudicates whether a takedown decision by Facebook or Instagram is consistent with the company's content policies.

Meta established the Oversight Board through an irrevocable trust that oversees $130 million in initial funding. It appointed twenty board members in mid-2020 and then another twenty members several months later. Its members have included former Danish Prime Minister Helle Thorning-Schmidt, former European Court of Human Rights judge András Sajó, and former US court of appeals judge Michael McConnell, as well as respected constitutional and human rights law experts, journalists, and internet activists. Under its founding charter, the board's decisions about specific content are final and binding.

The Oversight Board has not been shy about exercising that authority to reverse Meta's moderation decisions on hate speech, harassment, nudity, sexual activity, promotion of violence, deepfakes, and misinformation about COVID and other public health threats.

This effort, as well as the other moderation and related frictive design initiatives indicate that the platforms are keenly alert to consumers' distaste for certain kinds of content and content-distribution methods. This responsiveness is what the champions of the laissez-faire model wanted and expected. The First Amendment and Section 230 doctrines presume that companies will be better at driving out objectionable content than government regulators. Concern about losing users has arguably motivated platforms to be wary about some of the content that they promote.

But the companies are also fickle. Over the past couple of years, they have reversed course on many of these efforts. Meta, for example, announced that it would no longer amplify or demote posts about politics or political issues. In early 2024, it also shut down a celebrated social media monitoring tool that outside researchers and journalists used to track misinformation on Facebook and Instagram. Elon Musk, meanwhile, disbanded the teams responsible for developing frictive designs at Twitter (now X) soon after buying the company and rebranding it as a putative platform for free speech. The current laissez-faire doctrine gives the companies a pass on the decision to moderate less, even when informational harms are likely to follow.

Advertisers might have some impact on the safety of the content that appears on platforms. But (as we will see in the next chapter) they are insufficiently powerful in the ad-tech political economy. Perhaps they are also fickle. In 2019, for example, the World Federation of Advertisers worked with leading brands to create the Global Alliance for Responsible Media. The group spearheaded a major boycott of X in particular soon after Musk disbanded the company's trust and safety divisions. In November 2022, the alliance's members agreed to withhold advertisements on X, as well as any other platforms that traffic in harmful content. The alliance shut down, however, after X filed a lawsuit alleging that the boycott was a collusive conspiracy that violated the antitrust laws.

The Romance Is Over

Section 230 is probably the single most important reason that social media and other companies that distribute third-party content and consumers' personal data have thrived as they have. Companies have purported to be mere platforms for user-generated content and third-party information. It has not mattered that they know that the information they feature is unlawful—or even that there is a good chance that it is. Courts have refused to impose a duty to monitor antisocial or malicious third-party users. Nor have the courts required companies to implement filters or other safety measures that are known to protect consumers from informational harms. They have given little to no attention to the safe harbor for companies that provide access tools to parents. Nor has it mattered when a service encourages users to post awful or false information about someone else.

As upside down as the current doctrine is, there is something arguably more pernicious at work—an aspect of the doctrine that is not apparent in cases like those involving Match, Armslist, or the Experience Project, but that, more than anything else, explains the current state of affairs. In the next chapter, we will explore the ways in which companies have exploited the cyber-laissez-faire ethos to build services that, *by their design*, hold consumers' attention *in spite of* those very consumers' intentions. The liability shield has generated an extraordinarily lucrative opportunity for companies to extract value from user-generated content without ever having to pay for its lived consequences. This problem underscores how important it is to reject rosy tales about how these companies are platforms for speech and, instead, to focus on the ways in which their commercial practices systematically facilitate informational harms by design.

Cyberlaw

Tech Rules in the Shadows

The Long Shadow of the Liability Shield

The cyber-laissez-faire regulatory framework has allowed companies to ignore the likelihood that their online services endanger consumers. This flips Congress's stated reason for the liability shield on its head. This is bad enough. But, as we will see in this chapter, there is something even more troubling at work.

For the past two decades, companies have exploited their generous protection under Section 230 as well as the First Amendment to build services at the expense of the very users for whom they purport to provide those services. Shoshana Zuboff has called this "surveillance capitalism," an evocative term that highlights the ways in which the companies "widen and accelerate the inextricable cycle of engagement › extraction › prediction › revenue." This chapter builds on this account to show that the beguiling free speech tropes on which social media and its boosters rely elide the companies' extractive practices. Social media, it shows, is generally not simple speech platforms at all. Its recommender systems and enchanting interface features reflect its extractive industrial designs on consumers' attention and personal data. The companies deliberately keep consumers glued to rake in profit from advertisers, data brokers, and other firms. They evaluate success based on measures such as likes, comments, reposts, upvotes or downvotes, dwell time, and watch time. They do this at the expense of consumers, as well as of authentic third-party content providers like news organizations who see only a fraction of the traffic that social media gets.

The companies autoplay videos of commercial scams and kooky conspiracy theories about viruses, vaccines, and voting because that content is arresting or just plain shocking. They recommend short video clips of people asphyxiating themselves even if they can expect that preteens will give it a try. Messaging apps feature speed filters that unsurprisingly induce young adults to use them while driving perilously fast.

Most social media does not engage in these practices in order to hurt consumers. It designs features like these because, as we have seen, Congress and the courts gave social media companies license. It took just a few years for companies to refine the engagement business model. And policymakers have failed to muster the political will to change course. This chapter returns to the 1990s to illustrate just how companies developed their businesses in the dark shadows of the liability shield.

User Control as a Business Model

As we saw in chapter two, most consumers in the early 1990s perceived the internet to be a boutique curiosity for researchers and hobbyists. CompuServe and Prodigy helped to bring early retail adopters online in the mid- to late-1980s, of course.

But adoption would not really start to grow until the mid- to late-1990s with the introduction of dial-up services like AOL.

Launched in the late 1980s, AOL was far more graphically appealing than CompuServe, which had a command-line design. It also offered a far wider range of services than Prodigy. By 2000, it had twenty million subscribers.

Besides AOL, a new generation of online services like GeoCities and Tripod promoted themselves as platforms for user-generated content, inspired by the features of early online communities like Stewart Brand's WELL. These early social networking services were also important in promoting general internet adoption. Their implementation of the user-empowerment concept was, for many people, a welcome alternative to highly curated retail services like those of AOL. The business model that these user-focused services developed would set the stage for the social media services we now take for granted.

Take GeoCities. David Bohnett launched the service in 1994. It hosted virtual individual pages called "neighborhoods" that any subscriber or "homesteader" could easily create. These spatial metaphors were evocative. They engendered a sense of community and user buy-in. Each neighborhood would be defined by the content it hosted. Homesteaders, for example, could create a page in "Hollywood" if their interest was in media and entertainment; fitness pages would be in "Hot Springs"; "Wall Street" was for homesteaders whose focus was finance; and "Area 51" was for sci-fi pages. "Heartland," which was easily the largest of all of the communities, focused on "parenting, pets, and hometown values," with neighborhoods with names like "Plains, Meadows, Prairie, and Woods."

GeoCities' growth in the late 1990s was explosive, reaching about thirty-eight million subscribers at its peak. Unsurprisingly, most people who joined the service were much like the generation of subscribers on CompuServe and Prodigy before: white and suburban. And they had disposable income. But, even at its $4.95 per month subscription rate, GeoCities did not find a sustainable business model that attracted investors until 1997, when it introduced pop-up and banner ads. Advertisers proved to be keenly interested in a platform that could promote products and services to discrete, ready-made audiences defined by neighborhood. A couple years later, the company debuted an e-commerce platform called Marketplace, which included a GeoStore where users could buy GeoCities-branded merchandise.

GeoCities also developed tools that allowed community volunteers to moderate and take down content based on predefined company guidelines. This would empower homesteaders and their visitors to cultivate virtual neighborhoods to their own taste, which arguably made the pages more welcoming to like-minded people. Volunteers readily monitored for content that violated those rules. The company also introduced automated tools that scanned for impermissible content. Bohnett would later recount that, without the volunteer system and the moderation tools, the company would have probably been overrun by pornography and offensively violent language—and never have grown in the way that it did.

By 1998, GeoCities had evolved into a media and e-commerce giant. Bohnett took the company public in 1999 at a value of $86 million. The only bigger internet companies at the time were AOL and Yahoo. The latter acquired GeoCities a few months later for $3.6 billion, at the peak of the dot-com bubble.

The acquisition proved to be a costly mistake after Yahoo implemented a series of changes that undercut the feeling of user control. Among other things, it stopped assigning pages by neighborhood and changed GeoCities' terms of service to claim ownership of all content on user-created websites. In response, homesteaders and other users boycotted. Yahoo ultimately backed down, but, by that point, the damage had been done. In the years that followed, free webhosting siphoned off many former or would-be GeoCities users. In 2009, Yahoo shut down the service in the US for good.

User Engagement as a Business Model

A handful of websites filled the gap that GeoCities' demise left behind. They were not internet services in the same way GeoCities was. But they had the same user-control features that appealed to a growing segment of the consumer market. Importantly, and perhaps unwittingly, they relied on the legal protection that Congress enacted under Section 230 for online services.

For example, Friendster and MySpace, which became available to users in 2003, enabled ordinary users to customize their own public pages. Both attracted a lot of buzz. Tech writers marveled at the way in which Friendster in particular had conceived of a "new kind of Internet"—"one more about connecting people to people than people to websites." Its innovation was that it employed algorithms that estimated every new user's connection to existing users, giving the company a bird's-eye view into social groupings and networks. MySpace, meanwhile, promoted itself as a site through which musicians, artists, and organizations could cultivate fan bases and followers.

Friendster exploded in popularity in much of Asia, while MySpace proved especially attractive in the US. Within just a few years, the two companies each grew to about 115 million users. But then both fizzled out in the decade that followed. Friendster was done in by the massive amount of computational power that its network analysis required as well as by internal confusion about the brand's identity. For MySpace, it was arguably redesigns by News Corp., which acquired the company a couple years after its founding.

Reddit, which started in June 2005, also thrived. It afforded even more user control across its resident user-moderated pages. The platform hosted forums ("subreddits") in which "redditors" could submit, comment on, or vote on ("upvote" or "downvote") links, texts, images, and videos posted by others. The more popular a post, the higher it would appear on the site. The most popular ones would appear first on Reddit's home page. Subreddits coalesced users who shared an interest in a specific topic, which, as GeoCities proved, was a good way of generating advertiser revenue; advertisers are eager to reach ready-made and authentically user-generated consumer markets.

Reddit would never be among the top five social media sites but, largely due to its user-focused moderation format, it could credibly claim to be the "front page of the internet." Its impact on the culture more generally was also significant. It may not have invented them, but Reddit popularized durable online moderation conventions like upvoting/downvoting, AMAs ("Ask Me Anything") featuring celebrities and notable figures, and user comment syntax like DAE ("does anybody else . . . ?").

In October 2006, Condé Nast acquired Reddit for $20 million, and, by 2011, the company became an independent subsidiary of Condé Nast's parent company, Advance Publications. The company was valued at $500 million in 2014 and $10 billion by August 2021. As of May 2021, Reddit had 234 million monthly users. By 2024, it surpassed 500 million global accounts, with projections of over 556 million users by 2028. Reddit went public in March 2024.

Facebook, however, was easily the most commercially successful platform for user-generated content in the years following Section 230's enactment. The company's amazing rise is conventionally seen as a triumph for Silicon Valley and the ideological faith in the power of platforms for user-generated content. Today, two decades later, its origin story is the stuff of legend. Columbia Pictures made it into a feature film just six years after the social network's founding.

As legend has it, in 2004, Mark Zuckerberg and a couple of Harvard friends founded the service as an online student directory through which users could share personal updates, photos, and video. Facebook distinguished itself (back then and today) by serving what it perceives to be personally "relevant" content to each user through a "NewsFeed." Introduced in 2006, the NewsFeed is a perpetual scroll of friends' updates on each user's personal homepage. At the time, it felt novel and alive. As one columnist wrote a few years later, it made the service feel like it had "a soul."

A year later, the company added its famous "Like" button to each post through which users could convey enthusiasm. Facebook also displays the number of "Likes," transforming the act of "Liking" into a social event through which users express collective appreciation. Enough people could transform a single innocuous post into a viral one. Facebook would add other "reactions" in the decade to come.

But the "Like" button does much more. At the start, Facebook, like its peer social media services, published posts in reverse chronological order—the most recent would be first, followed by the next most recent, and so on. The "Like" button, along with browsing and search history, as well as other information Facebook collected about users, supplied the company with valuable information. With the help of predictive algorithmic models, the company could use this data to sequence and customize content for each user in real time, regardless of when friends posted it.

To be sure, Facebook was not the first company to introduce predictive algorithms in a consumer-facing service. Outside of social media, Google Search, YouTube (which Google acquired in 2006), and retail giant Amazon had been experimenting with and deploying personalized recommender systems and user-based collaborative filtering. The NewsFeed personalization algorithms, however, were not overtly transactional in the same way. Facebook used them to cultivate intimacy and social connection in ways that were not really the main focus for those other companies.

Some users resented personalization. They thought it was creepy and intrusive. But mostly it was unfamiliar. Targeting algorithms, after all, were relatively unknown to most users at the time. As far as they were concerned, the company might as well have had human operators who were monitoring users' every move online.

In time, however, automated personalization proved to be one of Facebook's most enduring innovations. On sites like Friendster and MySpace, a user's content would mostly sit unseen on a personal page. Through the NewsFeed, in contrast,

Facebook *served* posts to the people it predicted would most likely be interested. It would not take long for consumers to gravitate to Facebook, despite their initial reservations. The NewsFeed also inspired the company to develop algorithms that profiled and sorted users in ways that were far more granular and less expensive than was possible in the ad-driven print, television, or cable industries. It soon became the primary mechanism through which the company would hold and control consumer attention.

Over the years, Meta, Facebook's parent company (also owner of Instagram), has tweaked its content feeds and recommender systems across platforms to attend to its executives' shifting priorities and strategic concerns. In the mid-2010s, in response to concerns about the quality of content, for example, it began to weigh users' most recent interactions more heavily than older activity. In the late 2010s, it de-emphasized clickbait, inauthentic stories, and unsupported conspiracy theories and false claims about public health. It also experimented with prioritizing content from friends and other close connections, rather than strangers or organizations.

In early 2025, soon after President Trump's inauguration, Meta ended its fact-checking program across its platforms in response to concerns that it was part of the anticonservative "censorship industrial complex."

The company insists that this shift reflects a greater respect for free expression; it promises to leave it to users to correct or challenge each other's posts. This claim of devotion to free speech is hard to square with its unrelenting aim to personalize users' experiences. Those decisions have never really been users' to make. They have always been Facebook's.

As it has changed and grown over the years, Facebook has amassed an unrivaled global user base, which is at around 3.1 billion today. Of course, not all of Meta's ventures have been successful. It notoriously incurred $13.7 billion in losses in 2023, largely due to an ill-conceived foray into the "metaverse," a virtual reality platform for socializing and commerce. For the most part, Facebook has remained Meta's flagship. Today, given its size and reach, Facebook has become one of the most influential destinations on the internet.

Facebook's NewsFeed quickly inspired a wide range of surveillance-based social media across content types and market segments. For example, Pinterest, which launched in 2009, enables users to search, upload, curate, and share images and videos ("pins") across various interests like fashion, books, and home decor. It developed a "Home Feed" that displays followers' pins as well as suggested accounts and pins, and an "Explore Feed" that highlights trending ideas. In 2012, the platform also enabled companies to promote their products and services within feeds and alongside search results. This is the main way by which the company makes money: It targets ads and partners with influencers and merchants who pay commissions when users purchase products. Pinterest went public in April 2019 and now has 570 million monthly active users, most of whom are women. It has also built or acquired a wider range of features. For example, it added private direct messaging (Messenger), a platform through which users could buy or sell anything (Marketplace), a service that helped advertisers to target their brands to the right audiences (Ad Manager), and a popular platform through which users could play games with friends.

Other companies have applied the recommendation-and-engagement model to specific market segments and practices. LinkedIn, which officially launched in 2003, is the most popular networking platform for professionals, job seekers, recruiters, and potential clients. Its revenue streams are in job recruitment "Talent Solutions" for employers, premium subscriptions for professionals, and, of course, advertising to the coveted professional class of users. The company claimed its first year of profitability in 2006 when it integrated its "People You May Know" and "Recommendations" feeds. Microsoft acquired LinkedIn in 2016 for $26 billion and it has been among the software giant's most lucrative assets, generating about $15.1 billion in revenue with a user base of about 1.2 billion around the world. Most of its users are young professionals between the ages of twenty-five and thirty-four, one of the most coveted market segments.

Smartphones: Expanding and Deepening Engagement

Pinterest and LinkedIn are just two of a wide variety of companies and experiments that seized on the surveillance-driven business model. The smartphone made the model more attractive.

Apple's launch of the iPhone in 2007 and Google's Android mobile operating system a year later greatly enlarged the scope of what was possible. Smartphones, after all, are multimedia interfaces by which companies can reach even more consumers around the world through a wider variety of applications for communicating, socializing, and, importantly, commercial surveillance. Consumers would no longer need a desktop or laptop to use such services.

User review and rating sites like TripAdvisor and Yelp, which originally launched in 2000 and 2004, respectively, easily adapted to the location-based surveillance model that smartphones made possible. They, along with newer apps like FourSquare and OpenTable, could use location to personalize recommendations and ads for restaurants and other local businesses.

Location-based surveillance technologies transformed the social media market. WhatsApp, which launched in 2009, allows users to send texts and make voice and video calls, as well as share images, video, and documents. Users can send these to one person or to a large group of connections. Along with Telegram and WeChat later, WhatsApp also integrated communication features like messaging, voice and video calls, and even payment systems. Today, nearly three billion people use WhatsApp every month, making it the most popular messaging app around the world.

SnapChat, the ephemeral messaging and media sharing app, started in 2011. Through it, users send messages ("snaps") that are only available for a brief period of time, mostly controlled by the sender.

Snaps may include a photo or video that users can edit with filters, effects, text captions, and drawings. Snapchat generates much of its revenue through its Discover feature, introduced in January 2015. Discover distributes ad-supported short-form content from major publishers like CNN, ESPN, and BuzzFeed. The app also monetizes location-based "geofilters" with corporate sponsors like Gatorade, and movie studio trailers. As of February 2025, the platform reaches 453 million daily active users.

Instagram founders Kevin Systrom and Mike Krieger were quick studies of the engagement model. They launched the service on iPhones in October 2010 as a photo- and video-sharing app through which users could "like" and comment on others' media. Their idea was to combine visual content-sharing with location-based commercial surveillance techniques.

Instagram was an instant hit. By the end of 2010, the app had one million users. Two years later, when Facebook acquired it for $1 billion, it had twenty-seven million. Facebook had adopted a defensive strategy of acquiring startups who posed a competitive threat on smartphones. For this same reason, in 2014, it acquired WhatsApp for $19 billion. (Facebook's owners created Meta, the holding company, to account for all of these acquisitions, with Zuckerberg as CEO.)

Then there is Twitter, now X. Jack Dorsey founded the company in 2006, months before the commercial launch of the iPhone. His microblogging platform turned out to be superbly suited to smartphones. (The case for Twitter was clear after it was featured at the iconic South by Southwest Interactive conference in 2007.) Like Facebook, Twitter users could post, repost, and like content. But its 140-character limit invited users to shoot off hot takes and replies on a handheld device. (The company doubled the character limit about a decade later.) The format emboldened users to post remarks that were glib, snarky, and frivolous. This rawness fostered a sense of license and authenticity. Advertising was meant to be its principal source of revenue, although the company struggled to make a profit. Elon Musk, X's current owner, promised to find new revenue streams, but that effort has languished, largely because of the backlash against his brazen hard right turn.

Ad Tech Rules in the Shadows

It is no surprise that social media has proliferated given how lucrative the engagement model has been. The ambition of the entrepreneurs in this market is to find ways to leverage the intermediary market position between users on the one hand and businesses who pay handsomely for access to consumers on the other. Advertisers are the keenest of them, spending over $234 billion worldwide in 2024. Forecasters project that this will climb to over $400 billion by 2029. In response to a bewildered US senator's question in a 2018 hearing about how Facebook makes so much money, Zuckerberg said, "We run ads."

Publishers like Fox News and the *Washington Post*, social media like TikTok and Instagram, and search engines like Google and Bing all use techniques for tracking and profiling consumers based on personal information like browsing history, networked device, operating system, and location. The largest platforms operate or contract with "sell-side" brokers to promote advertising spots or "inventory" across their sites and services. Through this system, advertisers reach a wider range of consumers in real time across the internet than they would if they negotiated for ad placements one at a time. No matter who the players are, behavioral targeting has dominated the online ad business.

But it is hardly perfect. Research and reporting shows that it tends to promote low-quality products that are not really very responsive to what consumers actually want. A 2023 Carnegie Mellon study found, for example, that, while targeted ads feature products that are more relevant to consumers, they are "also more likely to be associated with lower quality vendors and higher product prices compared to competing alternatives" that consumers could find in an online search. In the wake of this study, tech reporter Julia Angwin found that an "anti-woke" company targeted ostensibly "hypermasculine" middle-aged men to receive ads for "razors that feel like someone is pulling your facial hair out with a tweezer one at a time." Apparently, the razor gets a 2.7 out of 5 rating. The men who buy them, however, "liked the product's political message more than the razor itself." Perhaps there is value in that kind of purchase, in spite of the reportedly poor quality of the razor. Anyway, ad tech promoted this product in spite of its ineffectiveness as a product.

The companies have also been caught in the act of enabling unlawful discrimination in advertising. In 2019, following on groundbreaking investigative journalism by Julia Angwin's team at *Politico*, Facebook settled lawsuits that alleged widespread discrimination based on gender, age, and race on the company's "Ad Manager." The platform had openly enabled advertisers to target audiences by including or excluding people based on those "protected characteristics" as well as other demographic categories. Google reportedly allowed employers and landlords to discriminate against nonbinary and transgender people. It pledged to crack down on the practice only after being alerted to it by journalists. These restrictive targeting practices are especially pernicious because their victims are never made aware that they have been excluded.

The ad tech market's opacity has also enabled fraud. Even when companies purport to be transparent about the effectiveness of their ad delivery practices, they sometimes provide faulty or incomplete data. Some platforms and supply-side brokers inflate or otherwise mischaracterize the effectiveness of their targeting techniques. Advertisers and publishers have expressed skepticism about the ways in which the large platforms set prices for impressions, but nevertheless continue to work with them because of the sheer scope and size of their potential consumer reach. They also typically charge lower prices than traditional media.

In 2016, Facebook reported that, for the two preceding years, it had overestimated an important video metric by 20 percent. The company's metric reflected the number of "views" of a video, which could be as short as three seconds, rather than the total time users actually spent watching the video, which is longer than three seconds. This miscalculation overstated the performance of video ads. This claim does not look like a simple mistake given that the company had been urging providers to "pivot to video." Something similar happened to LinkedIn in 2020, when it admitted to overcharging four hundred thousand advertisers for marketing campaigns on its site after an internal review revealed that it overstated video views and ad impressions. The company explained that it had mistakenly charged advertisers for video ads that played even when a user was not using their device or quickly scrolled by an ad that would autoplay even when out of view. The magnitude of the error was relatively small (under $25) for most advertisers, but it suggests a flaw that could have deeper consequences for others.

These mishaps are likely more pervasive than people know, given how shadowy the ad-tech side of the business is. Advertisers have been pushing back. In 2011, a group of them filed a class action lawsuit alleging that Google overcharged on its AdWords ad platform. That case settled for $100 million in 2025. And in 2023, another group of advertisers filed a class action suit against Alphabet, the owner of Google, for inflating the effectiveness of its TrueView ad program and then overcharging based on those misleading metrics. Here, the advertisers alleged, Google inflated the number of video views by including autoplay videos and those that were muted. They also allege that Google included videos that played on unlisted web pages.

The market has also been tightly controlled by just a handful of companies, with Google sitting atop of them all. In its antitrust case against Alphabet, Google's parent, the US Justice Department recently showed that the company (through its advertising units and programs) abuses its market power across the ad-tech market through its sell-side platform and its ad exchange, as well as in the market for buy-side tools for advertisers. The litigation noted, for example, that Google has 87 percent market share on the ad-selling side of the business, putting it in a position to charge a 20 percent commission on every dollar that advertisers pay—at the expense of website publishers. The company also retaliated against firms that sought to challenge its control over publishers' ad inventory.

The FTC has also brought a series of actions against data brokers who unfairly traffic in consumers' sensitive data. In December 2024, for example, the agency entered a settlement order with Mobilewalla, a company that collects consumers' de-anonymized personal data from "real-time bidding" exchanges. Its investigation revealed that, while purporting to clients that it collects consumers' information to help place ads, Mobilewalla actually built "audience segments" based on the personal information that it would then sell to third parties. What made matters worse is that, in analyzing location information, it built audience segments that revealed especially sensitive and politically fraught personal information like the identity of people who visit reproductive health clinics and attended protests after George Floyd's murder. Consumers, meanwhile, had no clue. Such practices, the commission alleged, were unfair because they exposed consumers "to potential discrimination, physical violence, emotional distress, and other harms—risks consumers could not avoid since most were unaware" of the practice.

Alternatives to Ad Tech

Many companies operate alternative business models. For example, subscription- or fee-based content services like those of *The Wall Street Journal* and *The Boston Globe* sometimes work well. The most popular consumer-facing businesses, however, have very little incentive to move off of their targeting practices because they bring so much near-term gain. Targeted ads also tend to be dramatically cheaper than anything that existed during the heyday of traditional broadcast television or print publishing. Thirty-second Super Bowl or World Series spots and aromatic full-page *Vogue* or *Vanity Fair* perfume ads were relatively blunt ways of reaching interested customers. Behavioral targeting, on the other hand, reaches exponentially more consumers who are likely to be interested, whether the ads are about butt-less pajamas, Jell-O mold recipes, or right-wing militia gear.

Some research suggests that "non-engagement" techniques might be effective at holding consumer attention. One recent paper co-authored by industry and independent researchers, for example, shows that optimizing for content quality measures like "authoritativeness" (as opposed to sensationalism) is likelier to keep consumers interested. Companies may not pursue such alternatives, however, if they are untested at the scale at which they might be profitable. The companies would have to devote resources and forgo reliable near-term profit to even experiment with those other models.

Contextual advertising techniques are another alternative. Under this approach, companies do not necessarily rely on the identity of a consumer to deliver an ad. Instead, they serve ads based on the consumer's immediate, real-time experience—that is, based on the content of the site that a user visits at any given time or the requests they make. For example, instead of delivering an advertisement for groceries based on a given consumer's browsing history, contextual models serve ads based on the content of the recipe or food item page that the consumer visits as well as, perhaps, their physical location at the time of the transaction. In any event, many companies that rely on such models also rely in a limited way on some form of individualized targeting. Phone location tracking, for example, is necessarily individualized.

Not What Section 230 Was For

The ad-tech political economy explains why platforms decline to moderate content in any way that resembles the original Good Samaritan logic for the liability shield. Even if companies once had such an incentive, they mostly lost it once the staggeringly lucrative business model came into focus. As soon as that happened, companies understood that they could, on the one hand, purport to be simple platforms for free speech and user-generated content and, at the same time, monetize consumer attention and personal information behind the scenes. The stickier their user interface designs, the better. Naturally, most companies' workings are not as sophisticated as those of large companies like Meta and Google. Armslist or the Experience

Project, for example, designed their services to induce a far narrower range of interactions. But these companies, too, fashioned their services with advertisers in mind.

Given this state of affairs, it should be plain as day that it is a categorical mistake to conceive of or even talk about social media as if it is a simple platform hosting user-generated content. To describe it that way obscures its pecuniary reasons for being and understates its dramatic systemic influence.

This is not hyperbole. Recall President Trump's exhortation to his social media followers in the wake of the 2020 presidential election to "stop the steal." Distribution algorithms at prominent services like Twitter, Facebook, and YouTube immediately amplified those false claims, before the companies blocked them later. But by then the damage had been done. Those posts galvanized the groups that violently ransacked the Capitol on January 6, 2021. The platforms were not necessarily in cahoots with the organizers of the attack, but their automated systems recognized an opportunity immediately. Unconcerned about the legal risks, content distribution algorithms accelerated the "stop the steal" meme to the people who were most likely to break into the Capitol Building in broad daylight.

Even before the 2020 election, whistleblowers had come forward to describe the lengths to which companies have gone to keep consumers glued, even in the face of resistance from within. Years after leaving his post as president of Facebook, for example, Sean Parker described himself as "something of a conscientious objector" when it comes to the engagement model. He expressed worry that the company was having unknown social impact, much of which was probably not good. This is because Facebook's aim, as with that of other social media, is to draw as much of users' "time and conscious attention as possible." He likened its moderation choices to controlled "dopamine hits" that exploit "a vulnerability in human psychology." Kids, he lamented, are the most vulnerable.

A few years later, very soon after the January 6 attack on the Capitol, former product manager Frances Haugen copied and published thousands of pages of internal documents showing how Facebook knowingly retained an amplification algorithm that demonstrably promoted violence, as well as content that diminished the self-esteem of young women. A more recent Meta defector, Sarah Wynn-Williams, the director of global policy at Facebook from 2011 to 2017, wrote a scathing tell-all book about the company's unmitigated ambition for consumer attention.

These accounts all clearly show that Facebook and other major social media companies are mainly in the business of designing services in much the same way casinos design slot machines. They are flashy and, for many people, hard to resist because of the way in which they constantly dispense "unpredictable rewards on schedules structured to exploit human psychology and get people hooked." This is why some researchers and writers use the language of gaming to describe platforms' penchant for publishing a post's total likes or reactions, reposts, and shares. These tallies resemble the leaderboards that you might find in a competitive video game or the feeling one gets after successfully pulling a slot machine lever in a casino. Users experience a visceral sense of pleasure (technically, a release of dopamine) when they feel that they have earned an unexpected reward from playing such games.

Kids are the easiest targets. The American Psychiatric Association recently warned that teens "may be particularly vulnerable to technological addiction because their brains are still developing." In this regard, a federal appeals judge who heard oral argument in a case concerning the constitutionality of California's Protecting Our Kids from Social Media Addiction Act wondered whether the companies may be designing their services in the same way that tobacco companies once marketed cigarettes to teens. There is something to this. At least, as social psychologist Jonathan Haidt's chart-topping book *The Anxious Generation* concludes, platforms seem to be fostering unprecedented levels of anxiety and depression among young people. According to a 2023 surgeon general's advisory on Social Media and Youth Mental Health, the current research ought to give policymakers and public health officials pause.

Legislators, judges, and advocates for the liability shield could not have anticipated any of this three or even two decades ago. They probably did not foresee the ways in which the immunity would create an irresistible incentive among platforms to hold consumer attention. Given their faith in self-regulation, they surely did not imagine that some platforms would be so motivated by this pecuniary incentive that they would become indifferent to the ways in which their services could be dangerous. Advocates back then really seemed to believe that the companies would always be simple platforms for free speech and user-generated content. They assumed, as the courts did in the *Zeran v. AOL* and in *Reno v. ACLU,* that all users are similarly situated virtual pamphleteers.

Even today, Big Tech advocates remain undeterred in their stated faith in the liability shield. Their commitment to the idea that today's internet continues "to be defined by user-created content" is plainly at odds with what we know about the prevalence and impact of the engagement model and ad tech. The information asymmetries between the major platforms and consumers are starker than they have ever been. A handful of internet companies engineer most consumers' online experiences through their user interface designs and content-targeting practices. They and their boosters brazenly seek to shield these techniques from legal or regulatory scrutiny on the fanciful theory that the companies are platforms for free speech and user-generated content.

Until Congress repeals or rewrites the statute, solving the problem will rest with courts. Over the years, judges have rarely evaluated the extent to which a defendant social media service is actually complicit in the violation of law. And neither plaintiffs nor the public ever get a chance to review the documentary evidence in these cases, since courts tend to give it legal protection from disclosure early in the litigation.

Fortunately, as we will see in the next chapter, things are changing. Courts and policymakers are warming to novel new design-related theories that are suited to our time—theories that scrutinize services for their specific design features rather than the content that they publish or distribute. This is an important development that portends more accountability for an industry that has gotten away with so much.

Holding Platforms Accountable

Seeing Platforms for What They Are

For nearly three decades now, retail consumer-facing services on the internet have avoided public accountability because they purport to be simple platforms for free speech and user-generated content. As we have seen, however, their takedown decisions, moderation techniques, and interface features reflect industrial objectives that are more or less indifferent to the information they convey. Their priority is holding people's attention. And this design thinking has sometimes come at the expense of their most vulnerable users. This reality is far different from the state of affairs in which Congress and the courts first set out the prevailing hands-off regulatory approach.

The legal protection that these tech companies enjoy is rare. The only companies that benefit from anything close are gun manufacturers. Nonprofit organizations, churches, and schools, as well as landlords, auto and drug manufacturers, retail superstores and fast-food chains, banks, and insurers, to name just a few, are all subject to some form of consumer-protection law and related regulatory oversight. Laws and regulations in these settings protect consumers from harms that they can rarely anticipate or mitigate alone. These interventions are particularly important given the power and information asymmetries between consumers and companies. The same should presumably be true for online services, where people generally can neither really understand the "Terms of Service" to which they feel obliged to agree nor mitigate against the risks of harm when they use the services.

State and Federal Regulation of Harmful Designs and Ad Tech

The courts have given the companies a lot of leeway over the past three or so decades. But, as we will see next, their mood has been changing. They are likely being pulled in this direction by the consumer advocates and parent groups that, as we saw in the last chapter, have been filing lawsuits. They have also been getting some guidance from state and federal regulators.

State regulators in particular have been setting their sights on harmful service design features and commercial surveillance practices. Their efforts vary significantly. Illinois has been an early leader, with its statutory protections against abuses of biometric data. Vermont was among the first to pass comprehensive protections from the data broker industry. Other strong state laws, like those of California, Colorado, Minnesota, and Oregon, establish individual rights to notice, access, correction, and deletion. Most also impose limitations on how companies collect, retain, or sell their residents' personal information, as well as some combination of disclosure requirements and impact assessment requirements. These laws generally give consumers the ability to opt out of targeting and profiling, and in some cases also require impact assessments for activities conducted by data controllers that have a heightened risk of harm. Many also impose restrictions on the companies' use of automated decision-making systems.

California stands out. It is the only state that has created a stand-alone Privacy Protection Agency, which has enforcement and rulemaking authority. It is also the only state to allow residents to file their own lawsuits for data breaches. All of the other states give the state attorney general the sole authority to enforce the respective laws. The more business-friendly laws, like those in Virginia and Iowa, contain several carve-outs for certain industries and do not require consumer consent for data practices.

Meanwhile, a handful of states—California, Connecticut, New York, and Utah—have also enacted statutes that aim to protect children from online harms. Here, too, California has been a leader. In 2022, drawing on ideas from Britain's consumer protection regulator, California legislators enacted the Age-Appropriate Design Code Act. The law requires companies to implement the highest level of privacy protection for children, conduct data protection impact assessments, limit or block targeted advertising to children, prohibit "dark patterns" in user interfaces that manipulate children into providing personal information or doing things that are "materially detrimental" to their well-being, and prohibit the collection of precise geographic location information. States as disparate as Utah and Vermont have since enacted similar laws.

The industry is of course challenging these laws as unconstitutional intrusions on the freedom of speech. They are winning some of these cases. But they are losing others. The variety in the decisions has been an occasion for learning. Over time, regulators across the states have been settling on a consensus about the subjects and forms of regulation through trial and error—what journalist Alfred Ng has called a "Goldilocks zone." At the top of this list are protections for children and bans on the sale of precise location data.

Federal regulators have also acted in the absence of meaningful legislation or guidance from Congress. The Federal Trade Commission has been a leader in this area. Even as appointed commissioners from both parties have recognized that legislation would be better at redressing commercial surveillance abuses, they have moved aggressively to level the power and information asymmetries that define the online market for consumer services. The agency relies principally on the same statutory prohibition on "unfair or deceptive acts or practices" that it uses to fine brick-and-mortar fraudsters. As with the European Commission and other consumer protection agencies around the world, it has been particularly keen on stamping out exploitive "dark patterns" on retail consumer interfaces. In this vein, it has invoked the Restore Online Shoppers' Confidence Act to sue Adobe for hiding the early termination fee for its most popular subscription plan and making it difficult for consumers to cancel their subscriptions.

In its most notable recent action, the FTC entered a $500 million settlement with Epic Games, the developer of the popular multiplayer game *Fortnite Battle Royale*, for duping children into making unintended in-game purchases without their parents' permission. (The FTC focused, for example, on how it takes one click on the game hand controller to make a purchase but several unintuitive steps to cancel one.) The agency has also challenged smaller companies' application designs, including anonymous chatrooms in which cyberbullying is rampant. The FTC fined NGL Labs $5 million for enabling users to share a link on social media that invited followers to answer the prompt: "If you could change anything about me, what would it be?" Predictably, the answers that teenagers submitted were often cruel and dehumanizing. This is a toxic mix for all adolescents who are looking for their place in the world. It is potentially fatal for young people who are struggling with mental illnesses like depression.

Recent successful FTC enforcement actions have also taken aim at ad tech, the engine of the engagement model. In December 2024, the agency entered a settlement order with Mobilewalla, a data broker that collects consumers' de-anonymized personal data from "real-time bidding" exchanges. (These exchanges are a key player in the ad-tech supply chain. The companies administer automated auctions to target ads at the speed of light to a consumer who is on a specific website or application. Not only do these companies have access to all the data that websites and applications collect about their users, including browsing history and precise geographic location, but they also share this sensitive personal information with advertisers who bid for ad placement.) The FTC's investigation revealed that Mobilewalla never actually used users' personal information to place ads. Instead, the company built "audience segments" based on this data, which it would then sell to third parties who, in turn, would use the information to target ads.

According to the FTC, Mobilewalla built an audience segment of pregnant women based on the location data of women who visited pregnancy centers. The company similarly created audience segments based on the information it had about people who attended protests in the wake of George Floyd's murder. Consumers had no clue that Mobilewalla (or probably any other player in this system) had access to their sensitive data. These practices, the commission alleged, were unfair because they exposed consumers "to potential discrimination, physical violence, emotional distress, and other harms—risks consumers could not avoid since most were unaware" of the practice. This case comes after a series of cases that the commission has brought against other data brokers who unfairly trafficked in consumers' sensitive data.

Refining First Amendment Doctrine: Scrutinizing One Commercial Application at a Time

Over the past couple of years, the courts have been catching up. Advocacy groups and plaintiffs have been advancing new legal theories about the harms that discrete commercial surveillance practices and design features cause, and judges are increasingly receptive. They are recognizing that the companies are not necessarily the simple platforms for free speech that the courts described in *Reno v. ACLU* and *Zeran v. AOL* in 1997. They are increasingly persuaded that companies engage in practices that are unrelated to speech as such. They recognize, moreover, that consumers cannot generally mitigate product features and defects before it is too late. They understand that the alluring concepts of user empowerment and user choice simply do not map on to the way things actually work.

Take for example the Supreme Court's decision in *Moody v. NetChoice* to return the industry's audacious challenges to Florida's and Texas's social media laws to the lower courts for more fact-finding. Some applications of the statutes could be constitutional, the Court wrote, if they address commercial practices that are unrelated to the companies' editorial judgment about content. Not all internet applications and functions, Justice Kagan explained in her opinion for the Court, are alike. Judges must be alert to the details of each internet application because some, like those for retail shopping or peer-to-peer payments or ridesharing, may not be as entitled to protection as newsfeeds or video streaming recommendations. They serve different functions and, accordingly, should be subject to differing levels of scrutiny—sometimes less than the laissez-faire approach had allowed. Justice Jackson, in a concurring opinion, was crystal clear in this regard: "Not every potential action taken by a social media company will qualify as expression protected under the First Amendment." The courts below, she wrote, must consider how the specific services "actually function" and the ways in which the state laws affect the companies' content-related practices.

Similarly, Justice Barrett wrote her own concurring opinion to observe that, for the purposes of evaluating whether the respective law impedes a company's editorial discretion, courts could very well evaluate purely automated newsfeed algorithms differently from those in which "a person is in the loop." She expressed doubt, for example, that a company's use of AI systems for personalizing content could always be entitled to First Amendment protection. This musing was not purely hypothetical. Remember Sewell Setzer, the fourteen-year-old who developed a codependency with a character he created on Character.AI with the platform's tools? In spring 2025, the federal judge in Florida hearing that case cited *Moody*, and Justice Barrett's concurrence in particular, to reject Google's argument that Sewell's

AI companion embodied constitutionally protected expressive conduct.

The Supreme Court's *TikTok v. Garland* opinion from January 2025 further illustrates the more nuanced emergent approach to social media legislation. The statute at issue there, the Protecting Americans from Foreign Adversary Controlled Applications Act, generally prohibits US companies from "distributing, maintaining, or updating" a "foreign adversary controlled application." TikTok, the popular social media app owned by Chinese tech company ByteDance, was always legislators' focus. The "TikTok ban," as the law is colloquially called, effectively requires TikTok to sever all material ties with its China-based owner. Legislators expressed concern about the ways in which a "foreign adversary" was allegedly using the app to collect information about US citizens and covertly distributing content to manipulate voters.

The company, along with creators and free speech advocates, challenged the law, arguing that it is a content-based regulation of speech. The Court in response held that the national security interest in protecting US residents from China's collection of personal data justifies the law. Legislators were focused on a narrow set of non-content-based geopolitical priorities; under that more forgiving standard, the statute was constitutionally valid.

Even as distinctive as the government's national security interest was in that case, the Court also noted that Congress's focus on the application's commercial surveillance practices behind the screen was not content-based. "Data collection and analysis," the Court observed, "is a common practice in this digital age."

For the past several years, the lower courts have been saying so as well. They, too, have become increasingly nuanced about platforms' commercial surveillance practices and design features. In August 2024, for example, a federal judge relied on *Moody* to reject NetChoice's First Amendment challenge to California's Age-Appropriate Design Code Act. The record, the court explained, did not indicate that a substantial number of the provisions of the state's new child online safety law, including those addressed to curbing harmful designs and "dark patterns" in user interfaces, were constitutionally invalid. The district court would have to do more fact-finding to allow the challenge to proceed.

Application Designs and Commercial Practices Are Not Always "Publishing"

The courts' increased sense of nuance with regard to online platforms affects its application of the liability shield. Courts are becoming less likely to describe online services as "mere conduits" for "political discourse" and "intellectual activity" in the same way the federal appeals court in *Zeran v. AOL* did in 1997. Today, it is beyond cavil that current online platforms look nothing like chatrooms on Prodigy or neighborhoods on GeoCities.

In 2023, the Supreme Court chose to look past Section 230 to evaluate whether, in two separate cases, *Gonzalez v. Google* and *Twitter v. Taamneh*, the biggest social media platforms were *actually* liable for terrorist attacks around the world. The relatives who brought the cases against YouTube, Twitter, and Facebook argued that the companies' recommender systems enabled the perpetrators to recruit and raise funds for the attacks. In the end, the Court determined that the companies had not aided and abetted violation of the antiterrorism statutes—without saying anything about Section 230. During oral argument, the justices were overtly frustrated about whether the liability shield could apply to the recommender systems in the cases before it.

Justice Thomas, who wrote the opinion for the Court, has in other cases invited advocates to bring a worthy case that would give him and his colleagues the opportunity to narrow the scope of the protection under Section 230. He made his most recent plea at the very end of the 2023–2024 term, where he dissented from the majority's decision to turn down an appeal involving Snap. Judicial intervention, he warned, is urgent. "There is danger in delay. Social-media platforms have increasingly used § 230 as a get-out-of-jail free card." On the one hand, the companies seek protection under the First Amendment when they "organize users' content into newsfeeds or other compilations." On the other hand, they seek cover under the liability shield to deny legal responsibility for the unlawful content that they feature. "In the platforms' world," Justice Thomas wrote, "they are fully responsible for their websites when it results in constitutional protections, but the moment that responsibility could lead to liability, they can disclaim any obligations and enjoy greater protections from suit than nearly any other industry." These inconsistencies and contortions call out for a course correction.

Platform Designs "Contribute" Content as a Matter of Course

Justice Thomas may very well get his wish; lower courts are now expressing real reservations about stretching Section 230 protection to apply to companies' design choices. The first notable opinion to signal this shift was the US Court of Appeals for the Ninth Circuit's 2008 decision in a case against Roommates, a website that matches potential co-tenants. There, the Fair Housing Council of San Fernando Valley alleged that two of the company's features violated antidiscrimination laws. The first was a drop-down menu of personal characteristics that all Roommates subscribers must complete. The list included characteristics (like age, gender, and sexual orientation) on which sellers and landlords may not rely in their advertisements about or decision to rent or sell housing. (Race was notably not among the categories in the list.) It also required subscribers to say which of those attributes they prefer in co-tenants. The company would use these preferences to make matches.

The Fair Housing Council alleged that, through its drop-down menu, Roommates facilitated violation of federal and state fair housing laws. It argued that the company had "created or developed" content within the meaning of the statute when it required people to convey protected information. The second feature that the council challenged was Roommates' "Additional Comments" section, where subscribers could voluntarily post information about themselves or prospective roommates. Subscribers used this space to freely express preferences about cleanliness and tidiness, as well as preferences for gender, race, or sexual orientation.

Roommates moved to dismiss the suit on Section 230 grounds. The court granted the motion as it related to the "Additional Comments" section of the website. The company, it explained, had no hand in providing the information that users shared in that section. Subscribers could communicate whatever they wanted, without direction from the company because the matches that it made were based entirely on third-party user-generated content. This, it explained, was precisely what Congress sought to encourage with Section 230.

But the Ninth Circuit rejected Roommates' claim to the liability shield for the prepopulated drop-down menu. The company, it held, was an "information content provider" within the meaning of the statute because the service "materially contributed" to the allegedly unlawful conduct by requiring all subscribers to choose from prohibited attributes. It did not matter that users actually chose an item on the list; they had no choice but to select a category that the law forbids. The federal and state laws forbid people from eliciting or communicating such information about a prospective renter or buyer.

Roommates is probably the second-most-cited Section 230 case after *Zeran*. Courts across the country have adopted its "material contribution" standard to evaluate whether a defendant is an "information content provider." It is handy because it differentiates between consumer-facing features that Section 230 protects (the "Additional Comments" section) and those that it does not (a menu of options from which all users must select). It seemed plausible back in 2008 that the Ninth Circuit's reasoning would open up a new way by which plaintiffs could overcome the liability shield.

But it would take time. In the near term, the Ninth Circuit's "material contribution" standard may have actually hardened courts against a more nuanced understanding of what it means to be an "information content development" under Section 230. One of the more remarkable cases in this regard involved a website, The Dirty, that openly encouraged its visitors to post false and damaging rumors about young women. One victim sued when the site urged people to share photos and salacious rumors about her personal life—and then posted that material on the website. The Sixth Circuit, which sits in Cincinnati, Ohio, held that Section 230 shielded the website from liability. It did not matter that the publisher cruelly invited people to share salacious information about the plaintiff. Nor did it matter that the company leveraged that content to extort her or other victims. The only question that mattered for the court was whether The Dirty "materially contributed" to the content. The specific offending content came from third parties, it held, not The Dirty. The Sixth Circuit cited *Roommates* for support.

Similarly, in 2019, the Second Circuit, which sits in New York City, dismissed a product liability case against Grindr, the hookup app popular with gay men. Matthew Herrick, the plaintiff, alleged that his former boyfriend created fake profiles in Herrick's name on Grindr and directed prospective suitors to his home and workplace. The service, Herrick argued, distributed his location information in ways that exposed him to harassment and threats of sexual assault. He brought claims for, among other things, products liability and negligent design. Moreover, the company, he alleged, did not implement safety features to protect him from the reasonably foreseeable harassment that he endured.

In anticipation of the Section 230 argument, Herrick argued that Grindr was an "information content provider" when it distributed his location information to suitors. The Second Circuit rejected the argument, holding that a third party, Herrick's ex-boyfriend, provided the location information. The court did not blink at the idea that the location information that Herrick's ex-boyfriend shared was "content" for the purposes of Section 230. It likened the location sharing feature to the "Additional Comments" section on *Roommates*; Grindr, it observed, merely provided "neutral assistance in the form of tools and functionality equally available to bad actors and the app's intended users."

Algorithmic Amplification May Not Be "Publishing"

A few months later, the Second Circuit dismissed another suit, this one against Facebook, for making friend recommendations. In that case, *Force v. Facebook*, families and estates of victims of terrorist attacks in Israel alleged that the social media company materially supported Hamas operatives by recommending hateful anti-Israel content and connections with terrorists. The Second Circuit in *Force* sided with Facebook. The company, it explained, merely made those potential matches "more visible, available, and usable." This, it reasoned, is "an essential part of traditional publishing" and, therefore, is shielded under Section 230. The court repeatedly relied on *Roommates* to reach this conclusion.

The decision in *Force* was not a surprise. The case nevertheless is striking because the late Chief Judge Robert Katzmann wrote a separate opinion in which he disagreed with the majority's analysis of "friend- and content-suggestion algorithms." Neither the statutory text nor legislators' justifications for it in 1996, he reasoned, supported the view that Section 230 categorically shielded companies from harms arising from their algorithmic recommendations. He would have held instead that Facebook's targeted recommendations should not count as "publishing" because they created "real-world (if digital) connections" with demonstrably real-world consequences.

Even as a dissent, Judge Katzmann's opinion was a watershed. The opinion has quietly moved advocates, policymakers, and other judges to reconsider the liability shield given the ways in which platforms shape consumers' online experiences. Even members of the Ninth Circuit, which is mostly responsible for the broad reading of Section 230, have admiringly discussed Chief Judge Katzmann's discussion of recommender systems. Justice Clarence Thomas, who was Judge Katzmann's ideological opposite, cited the *Force* dissent approvingly in another case. "It strains the English language," Justice Thomas wrote there, "to say that" Facebook is "publishing" when it selectively recommends content to users.

Refining Section 230 Doctrine: Scrutinizing One Design Feature at a Time

It is no surprise, therefore, that the Katzmann dissent has inspired plaintiff attorneys, consumer advocates, and civil rights groups to bring cases against companies based on their service designs, well beyond promotion algorithms. Autoplay (where video plays without the user's prodding), infinite scrolls, and randomized matching, for example, do not resemble "publishing" in any contemporary or historical sense—or within the meaning of the statutory liability shield.

The most impactful decision in this vein came two years later in a case involving Snap, the popular ephemeral messaging app. In *Lemmon v. Snap*, the Ninth Circuit, in rejecting the Section 230 defense, allowed surviving parents to show that one of the application's ostensibly playful "filters" was responsible for the wrongful death of their young adult sons. Justin Davis, who was driving, rammed his car into a tree at 113 miles per hour when they were using the app's "Speed Filter," a real-time speedometer on their phones. According to the parents, Snap (the owner of Snapchat) knew that users went faster than 100 miles per hour on the belief that the app would reward them with in-app "trophies." Snap did not do anything to dissuade their users from believing that the perk did not actually exist.

The company moved to dismiss the case, arguing that the suit sought to impose liability for publishing user-generated content—here, perversely, the young men's driving speed. The Ninth Circuit rejected Snap's defense because the parents' claims concerned the design of the application, not Snap as a publisher. They alleged that the "Speed Filter and reward system worked together to encourage users to drive at dangerous speeds." It did not matter, the judges concluded, that the company provided "neutral tools," as long as the plaintiffs' allegations were not addressed to the content that users generated with those tools.

Lemmon has made an impression: The Ninth Circuit categorically rejected the Section 230 defense. Emboldened by the court's reasoning, consumers and their advocates have brought other design-related claims against platforms. Sometimes the Section 230 defense prevails, as in a Connecticut state case in which a mother alleged that two adult men on two different occasions blackmailed her teenage daughter after soliciting compromising photos from her on Instagram. The court there held that matching algorithms are shielded under Section 230. But, more than ever before, these cases are overcoming the liability shield.

A federal trial court in Oregon, for example, relied on the Ninth Circuit's reasoning in a sextortion case against Omegle, a web-based free chat service that randomly matched users in one-on-one chatrooms. The plaintiff, who goes by "A.M.," sought to hold Omegle accountable for pairing her with an adult stranger when she was just eleven years old. On the latter's request, A.M. had sent him pornographic pictures and videos of herself and her prepubescent friends. He then threatened to distribute the videos if she did not share more. The service did not require users to register or convey any information about who they were, including their age. When she became an adult, A.M. filed the suit, alleging that Omegle was defective under product liability and negligent design theories.

Omegle argued that it had no duty either to monitor the interactions between its users (even when a child was paired with an adult) or to warn people about potentially dangerous matches. But the district court rejected the company's claim to immunity. Omegle, it found, could be held responsible for matching A.M. with a sexual predator in his thirties. Omegle shut down its chatroom matching service a few months after the district court's decision—and after dozens of other families and young victims had come forward with similarly horrifying stories.

Parents, school districts, and child advocacy organizations have been carefully crafting their complaints to go after the ways in which platforms design dangerously addictive services. One of the most prominent is a case in the Northern District of California that consolidates hundreds of federal suits from across the United States against Facebook, Instagram, YouTube, TikTok, and Snapchat. The case addresses its allegations to the services' addictive designs. It argues that the platforms aim to promote "addictive and compulsive use" without concern for the substance of their content. Features like "For You" feeds, infinite scrolls, and autoplay mean to prolong use. Kids are especially susceptible, but these design tricks entrap everyone. The companies know this but nevertheless fail to impose screen time limitations or other safety features, at least in the US. (In China, social media must abide by screen time limitations and bans on late-night push notifications.) They do not warn parents of the risks of addiction.

Many of the plaintiffs' design-focused claims have been prevailing over the Section 230 defense. California District Court Judge Yvonne Gonzalez Rogers explained that the liability shield analysis requires a nuanced understanding of the service features. The doctrine blocks plaintiffs from suing platforms for matching users based on the content those users share. But services that recommend adults to children *before those users have shared content* are fair game; such claims are squarely about the design of the platform. Similarly, plaintiffs' claims that attack the companies for failing to implement robust age-verification or options for self-restricting screen time are not equivalent to claims about speaking or publishing. Judge Rogers has also allowed plaintiffs to pursue their claim that the companies pace and cluster notifications to induce addiction. Meta even owns a patent for a system that determines the best times to send notifications so that consumers are likely to respond. These features, Judge Rogers explained, are the companies' service features, not "publishing."

One of the more notable cases that illustrate the new mood among judges is Nylah Anderson's. Remember: She is the ten-year-old girl who died of asphyxiation after watching a series of "blackout challenge" videos that TikTok recommended to her on her For You page. Tawainna Anderson, Nylah's mother, sued TikTok for products liability, negligence, and wrongful death. Her complaint made clear that she was not seeking to hold the company responsible for content that originated from the reckless young people who were posting these clips. Her focus was on TikTok's decision to populate Nylah's individualized feed with blackout challenge videos.

The Third Circuit agreed with Anderson and rejected TikTok's Section 230 defense. The most notable aspect of the opinion is the way in which Judge Patty Shwartz, writing for a panel of judges, wrote that when TikTok recommended the blackout challenge meme to Nylah, it was engaging in "their own expressive activity or content (i.e., first-party speech)." This, she wrote, was its First Amendment right. However, Judge Shwartz continued, the companies cannot then turn around and disclaim responsibility under Section 230 for amplifying some forms of content over others. Recommender systems are "first-party speech" for which the companies should be held accountable.

Cyberlaw Reforms, Same Old Story

State policymakers, federal agencies, and courts are now more alert than ever to the threats and challenges that social media and other online services pose to consumers. Congress, however, has failed to recalibrate the regulatory approach that it put in motion three decades ago. Over the past few years, leaders in the Senate and the House have come close to passing comprehensive data protection bills and, to a lesser extent, major revisions to Section 230. Those proposals failed, however, mostly due to substantial pushback from the industry and from staunch free speech advocates.

This is not to say that Congress has been completely silent. In 2018, it enacted an exception to the liability shield for sex-trafficking. And in spring 2025, it passed the Take It Down Act, a law that criminalizes the nonconsensual publication of intimate visual depictions, including AI-generated deepfakes. (It seems that nothing motivates legislators more than protecting kids and criminalizing sex-related harms.) The law also imposes a notice-and-takedown regime on platforms, requiring covered platforms to take down a deepfake within two days of having notice of it.

The problems that these reforms seek to redress are merely the tip of the iceberg. Social media and other online consumer services cause or facilitate harms beyond those related to sex-trafficking and deepfakes. Nor do they just involve kids. Nor are they limited to misinformation, disinformation, polarization, bigotry, and misogyny.

Those are surely bad enough. But, as we have seen, online consumer services cause a wide array of deeply consequential consumer harms, including deceptive designs and dark patterns that dupe people into giving personal data or buying things they do not really want. Companies deliver ads for high-stakes opportunities and services like housing, employment, and healthcare that discriminate on the basis of gender, age, and race. They employ service features and sycophantic AI models that seize on consumers' deepest vulnerabilities and push them into feelings of self-doubt and depression. Sometimes these online services drive consumers to self-harm.

Online services do all of these things because they draw and hold consumer attention, an opportunity for which advertisers pay handsomely. Even in light of the doctrinal refinements we have seen over the past few years, the law has done nothing to slow any of this. To the contrary, the current regulatory approach gives the companies every incentive to do almost anything to keep consumers coming back.

What, if anything, can policymakers do to make platforms care more about foreseeable harms to consumers and to the greater information environment? This book recommends that we look beyond the content that brings people to online platforms and, instead, at the incentives that drive the companies to design their services. This would be a far more prudent and enduring strategy. And it would avoid the land mines in the doctrine for content-based regulation.

This shift in focus will not instantly make the online information environment less polarized. Nor will it necessarily force out all disinformation, bigotry, discrimination, and other consumer harms. But carefully crafted reforms addressed to the companies' commercial incentives, practices, and design choices would likely have those salutary effects.

In the final chapter and conclusion, I first review how some reformers—mostly on the right—are doubling down on the very ideas that brought us to this point. Their proposals underscore the enduring appeal of regulatory laissez-faire. Meanwhile, more tech-focused reformers have promoted technologies and protocols that they argue will diminish or altogether dismantle the control that the industry exerts over users' online experiences. These reforms would have policymakers return to the internet's foundational norms of user control and decentralization. Advocates for these changes believe that companies have corrupted the anti-authoritarian ethos espoused by the pioneer generation.

But, as I have shown, these interventions place too much faith in the beneficence of the companies and technologists that provide online services and manage consumer data. The central problem has never been a lack of forbearance toward entrepreneurial or technological creativity. It has been policymakers' failure to recognize platforms and other internet companies for what they are: ordinary commercial enterprises whose bottom line often conflicts with the best interests of consumers. In this light, government must play a far more active role than the prevailing approach has allowed.

Doubling Down on Laissez-Faire

The new liability shield exception for sex trafficking and the criminalization of the distribution of nonconsensual intimate visual depictions could have been a preview for more general protections for all consumers. In 2026, however, momentum seems to be going in the opposite direction. This was as plain as day when Big Tech executives appeared on stage with President Trump on Inauguration Day. They, along with businesses of all kinds, had donated a record-breaking $239 million to the planning of the event that day. The question was not whether the new administration would attend to the plutocrats alongside him that day. The question was when.

Four months later, the House approved a budget bill, which the White House guilelessly called the "One Big Beautiful Bill," that would impose a ten-year moratorium on states and municipalities that implemented their own AI regulations. The Senate's revision of the bill would have withheld funding to governments that did so.

The law would have covered "any computational process" that could "materially influence or replace human decision-making." This wording is broad enough to encompass any use of modern computational technologies in practically any context, including those that officials use for criminal sentencing and public school placement, as well as those that protect consumers, including kids, from unfair decisions in high-stakes contexts like insurance, consumer finance, employment, education, and healthcare. The bill affected the forty-seven states that have either enacted such laws or are considering such proposals. In the end, Congress did not include the moratorium after substantial pushback from the states and consumer advocates.

That the moratorium was so close to passage is worrisome, but not all that surprising given the appeal of the laissez-faire approach. The most common justification for it was that it was three decades ago for Section 230 and the other cyberlaws. The president and many of his allies in Silicon Valley, including leaders at Amazon, Google, Microsoft, and Meta, have been eager to lift regulatory burdens on innovation. Sam Altman, the chief executive of OpenAI, testified to the Senate that many state laws are "disastrous" for AI development. The new law, he explained, would simplify compliance for AI developers across the country. According to another prominent argument for the proposal, state legislatures might be well-meaning but their AI policymaking will undermine "the nation's efforts to stay at the cutting edge of AI innovation at a critical moment when competition with China for global AI supremacy is intensifying."

On hearing these arguments for the moratorium, someone new to US policymaking might have resolved that state legislative activity involving any technology (let alone any subject) is an anomaly or a bad idea. But of course this is wrong. The country's founders expected that the states would make policy. The Constitution's Tenth Amendment explicitly leaves to the states "all powers" that are not otherwise delegated to the federal government. Such powers touch a wide variety of bread-and-butter issues and problems in public life.

Nearly a century ago, Justice Louis Brandeis extolled the states as the "laboratories of experimentation" in a highly cited dissent in a case involving an Oklahoma licensing requirement for electric refrigerators. Over time, he argued, states discover that some regulatory approaches work better than others, which inevitably redounds to the benefit of policymakers everywhere, including in the federal government. In the 1990s, the Tenth Amendment was at the core of the reform agenda of the Federalist Society, the most powerful organization of right-leaning activist lawyers in the country. For it, the states are an important check against federal power, especially in the context of high-tech industries like telecommunications. States have been instrumental in regulating high-stakes technologies, including nuclear facilities and electric car manufacturing, as well as telecommunications. Even today, members of the president's political party are concerned about incursions on states' rights.

With the moratorium in place, consumers would no longer have a way to hold companies responsible for things as varied as employment discrimination, surveillance pricing, tenant screening, or credit scoring. These are all areas in which states have intervened on clear evidence that AI and other automated decision-making systems have been demonstrably biased against Blacks, Latinos, people with disabilities, and low income people (who, when it comes to finance and health insurance decisions, often carry past-due medical debt). Rather than honor or build on the states' efforts to curb the foreseeable dangers of the newest information technology, Congress was prepared to double down on the laissez-faire logic that has sustained social media and other consumer-facing online services.

Notably, this line of thinking has also shaped the current administration's stance toward the EU's Digital Services Act, which imposes obligations on the largest social media platforms and search engines to address the systemic risks of their services. The current FTC chair has even warned companies against complying, arguing that the law is inconsistent with the First Amendment.

Returning to End User Tools

Given the current laissez-faire mood in Washington, DC, tech-oriented reformers elsewhere have sought other means of curbing the excesses of the big social media platforms. Some in this group draw on Robert Putnam's lament in *Bowling Alone* three decades ago about the demise of physical meeting places like volunteer association halls and bowling alleys, and seek to foster purpose-driven local discussions about topics such as local politics, parenting advice, product recommendations, and news.

One project in this vein is Front Porch Forum, a Vermont-based online social network that since 2006 has hosted text-driven forums dedicated to specific communities across New England and parts of New York. Its popularity grew in 2011, when it played a leading role in supporting communities after a major flood, and then again during the COVID pandemic. The forum has distinguished itself by scoffing at the features that power large social media. It does not, for example, have a real-time personalized feed or a Like button. And only users who demonstrably live or work in the identified location may participate in discussions or post any comments to a specific forum; users cannot reach people beyond their local community.

The site is free to individual members, but it generates revenue through local advertising, donations, and a premium service with a few more options. This service works because of a devoted cadre of employees or volunteers who read every post and moderate all discussions. For the forum, its service is "more like a corner pub. If a patron starts making a ruckus, moderators ask him to tone it down—then toss him out if he doesn't comply."

Another overlapping group of advocates also rejects the customer service model on which the major online services rely but, unlike the Front Porch Forum, embraces online moderation features that prioritize user governance, like upvoting and downvoting. They recommend features like X's "Community Notes," which allows a user to suggest context for any given post that they find misleading, false, or even harmful. Other users may rate the additional note as helpful. If enough people with diverse perspectives approve, that note will appear alongside the original post in question. This is one moderation feature that Musk, X's current owner, has retained on the premise that users are better at governing content on the platform than "censors" at the company. As X explains it, "we believe giving people a voice to make these choices together is a fair and effective way to add information that helps people stay better informed." This is obviously a different and far more scalable form of moderation than the approach at the Front Porch Forum.

Other reformers advocate cross-platform user tools that afford consumers greater control over their platform experiences and recommendation systems. One prominent example are browser extensions that empower consumers to personalize their social media feeds no matter which platforms they use. Such tools work in much the same way that browser plug-ins for personalized privacy protection do. Recent proposals also foresee consumers delegating content personalization to a trusted vendor or "agentic" AI-powered software to browse, evaluate, filter, sort, and format content from platforms across the internet, whether or not the platforms allow it. These tech tools could disrupt the ways in which companies measure engagement and serve ads.

Communications scholar Ethan Zuckerman recently filed a lawsuit against Facebook that sought to block the company from interfering with his ability to build or use "Unfollow Everything 2.0," a browser extension that disables the NewsFeed. (Meta has threatened to sue people and researchers who use similar tools.) To make his case, he relied on Section 230(c)(2)(B), one of the lesser known provisions of the liability shield that immunizes the use of technical means for filtering offensive user-generated content. A trial court dismissed the case because no one has actually used the tool, let alone been sued by Meta.

Conceptually, browser extensions like "Unfollow Everything 2.0" as well as more personalized agentic AI could help individual users reclaim a modicum of control over their online experiences. But such tools are definitely not for everyone; they can only be as effective as tech-focused researchers like Zuckerman make them. Even if many consumers could effectively manage and update their online experiences, as it appears many aspire to do, such extensions would amount to a pittance—literally at the margins of the network— because they would not redress the systemic information asymmetry that defines the relationship between most consumers and companies.

Decentralizing

Middleware is a more far-reaching fix that has gained traction among reformers. It would allow consumers on one social networking platform to post and receive information from friends, family, and acquaintances who primarily use other platforms.

Such cross-platform interactions could work if participating developers adopt a common technical protocol for interoperability within a "federated universe" of platforms. The model is akin to the protocols that already allow people to exchange emails across providers like Google and Apple, or to send text messages across carriers like T-Mobile and AT&T.

This approach would allow individuals to curate their entire online experience across different platforms and devices to their taste—what leading advocates call a composable user experience. While advocates recognize that middleware might not prevent bad actors from converging in their own echo chambers, they mostly have faith that it would "dramatically dilute the power of the platforms to amplify fringe views and take them mainstream." The feature that the underlying platforms would share in common is their commitment to the shared, open, and interoperable protocol. Today, ActivityPub, for example, is among the most notable of such protocols among a new crop of social networking platforms, including microblogging services Mastodon and Pleroma and video sharing site PeerTube.

Advocates for this approach draw their inspiration from the internet's foundational norms. As one technology writer explains, middleware would "push the power and decision-making out to the ends of the network, rather than keeping it centralized among a small group of very powerful companies." It "opens up space for chaotic, exciting new innovations, and erodes the high walls that monopolies build to protect themselves."

Most proponents of middleware agree that it is a next-best option. They recognize that legislators would have to do more, including mandating legally enforceable interoperability standards. Without such interventions, most ordinary consumers will continue to have very little control over their algorithmic content feeds or the ways in which online services exploit their personal data, whether they rely directly on a major platform, a trusted independent vendor, or agentic AI.

It is not obvious that such laws would survive a First Amendment challenge if courts find that they compel social media to engage in speech with which they do not agree—that is, that they inhibit the platforms' speech rights. Even if such challenges fail, moreover, widespread adoption of middleware or related interoperability provisions would not necessarily reduce the market power that the largest online services hold. We should probably assume that the big platforms will continue to strive to hold their users' attention as much as they can.

There is a further, more fundamental problem with the idea that middleware can correct the excesses of social media platforms. Without new, additional consumer protection laws and changes to Section 230, middleware may become just another opportunity for mischief at the expense of consumers. In other words, it is not a next-best solution without the very protections and interventions consumers have needed all along.

As we have seen, companies that agree to abide by standards or protocols one minute may unilaterally back out of their promises in the next. Recall what happened to GeoCities and MySpace after Yahoo and NewsCorp, respectively, acquired them. Or consider the content moderation changes over the past couple of years at Twitter—now X—and Meta. They both once proudly held themselves out as services that fostered safe and healthy online conversations, only to reverse course a couple years later. A dozen years ago, in the wake of the Obama election and the

 Arab Spring, people celebrated these companies as champions of democracy. More recently they have facilitated democratic backsliding. The companies have made these recent changes pursuant to the whims of their leaders: that is, due to deeply held ideological commitments (in the case of X), political cravenness (in the case of Meta), or simply because their business priorities changed.

Even with middleware, browser extensions, and other tech tools at their disposal, most consumers are at a dramatic disadvantage when it comes to online services. They cannot avoid or mitigate risks about which they know little. Often they have no choice but to accept moderation and surveillance practices for essential services, like an ed-tech application to do homework or a fertility-tracking application that holds valuable health history. For all practical purposes, most online services are simply black boxes to which most consumers feel beholden. At the most prosaic level, many consumers feel compelled to use an app simply because their schools, healthcare providers, employers, friend groups, or families do. These are systemic problems that call out for economy-wide public law interventions, not contingent protocols or end user tech tools. To be enduring, reform requires enforceable consumer protection rules.

Conclusion

Tech for People: Changing Incentives

It will not be enough simply to reject the laissez-faire mindset. The harder task will be for policymakers to enact and enforce laws that recalibrate the incentives driving companies to design services despite their foreseeable consumer harms and social costs. Yet Congress today appears paralyzed—or cravenly beholden to the president—making meaningful legislative reform unlikely anytime soon. Still, I outline here several ways legislators could curb the incentives that shape the current media environment. None would be sufficient on their own; commercial practices across the broader political economy all require reform. Such is the depth of laissez-faire's reach. Only by attending to these multiple fixes can we hope to improve the quality of the information that consumers post, see, and share.

A. Revisiting the Liability Shield

1. Repeal

Congress could repeal or sunset Section 230 altogether. This would be a dramatic change because, once in place, it would very likely open the floodgates of litigation against companies that traffic in user-generated content. Companies might become warier of hosting or distributing anything but the blandest or safest content. Smaller companies and startups might be especially unlikely to push boundaries since they tend to have smaller reserves dedicated to litigation defense. Repeal might also have the perverse effect of further consolidating the dominance of the largest companies.

Congress could make exceptions for all but the largest online services, but this would allow dangerous or harmful business models to spring up. The point of Section 230 reform, at least as I have suggested throughout this book, is not to promote innovation or competition but, rather, to protect consumers from harm.

Nevertheless, there is good reason to doubt that a total repeal would be cataclysmic, given the protection that companies already enjoy under the First Amendment for blocking or downgrading third-party content. (One of the less cited provisions of the statute—Section 230(c)(2)(i), which protects "good faith" removal or blocking of third-party content—really just restates this protection.) Repealing Section 230 would really only remove an obstacle to litigation discovery, a source of information that is otherwise available to defendants in all other industries.

Finally, a sunset provision—say, where the law would no longer be in effect three or five years later—could create a very strong incentive on the part of Congress to revise the law. But such a measure would really postpone the question to an uncertain future. Legislators would still have to decide how to improve the statute.

2. Lifting the Protection for Distributing Unlawful Conduct or Content

Congress would probably do better by explicitly narrowing the protection companies enjoy. It could do this in a handful of ways. It could, as Danielle Citron and Benjamin Wittes have argued, preserve the protection only for services that take reasonable steps to address online abuse and harm, including implementing "reasonable content moderation" guidelines. This would return the doctrine to its focus on "good Samaritans" and exclude bad actors. Alternatively, an amendment could, as Mary Anne Franks has advocated, make plain that the statute shields protected speech rather than "information." She would also like the law to lift protection for companies that are deliberately indifferent to unlawful content.

Or Congress could specify circumstances under which companies would never be able to invoke Section 230 protection. These could include, for example, cases in which plaintiffs seek injunctive relief—a non-monetary remedy that would simply stop a company from continuing to engage in the allegedly harmful practice. Congress could also create an exception for civil law enforcement actions brought by federal agencies or state attorneys general. The logic here is that the protection should focus on restraining litigious plaintiff attorneys, not public servants who have been charged with the responsibility of protecting members of the public and competition.

Congress could entertain exceptions for specific claims. Citron, for example, has argued for statutory exemptions for allegations of "privacy violations, cyberstalking, or cyber harassment." I have argued that the protection should not block the full sweep of consumer protection laws, including civil rights laws and rules against unfair or deceptive trade practices. Such an approach risks being underinclusive by establishing a normative hierarchy of harms that leaves other harmful conduct to be addressed another day. This potential problem is made worse because Congress, which is notoriously unable to act these days, may have to update the law from time to time. On the other hand, such updates have always been federal legislators' responsibility across other subject areas, including, for example, in tax and copyright.

3. *Lifting the Protection for Service Designs*

Congress could also make clear that companies may not seek cover under Section 230 in cases in which plaintiffs challenge a company's commercial practices and service designs. The wording of this revision could be tricky because, since 1996, courts have insulated some design features like matching and recommendations from liability. As legal scholar James Grimmelman has warned, it will be hard to crisply define which practices are worthy of protection and which are not. Indeed, it is likely that some of these designs are sufficiently innocuous to warrant some protection. However, others, like randomly matching a child with a child predator, clearly enable egregious harms that would not otherwise be possible.

Narrowing Section 230 protection would allow the courts to actually scrutinize the merits of lawsuits challenging the platforms' service design.

All eyes right now are on the trial court in Northern California that is reviewing hundreds of federal claims from across the United States against Facebook, Instagram, YouTube, TikTok, and Snapchat. The plaintiffs address their allegations to the services' addictive designs, including algorithmic prioritization of harmful content, infinite scrolling, autoplay, push notifications, and filters. They argue that the platforms aim to promote "addictive and compulsive use" without concern for the substance of the content. Kids are especially susceptible, but these design features seduce everyone. The companies know this but nevertheless fail to impose screen time limitations or other safety features—measures they are required to use in China and elsewhere.

For now, many, but not all, of the plaintiffs' design-focused claims seem to be prevailing over the Section 230 defense. District Court Judge Yvonne Gonzalez Rogers has ruled, for example, that services that match adults to children in chatrooms or through recommendations before any of those users has shared content are fair game; such claims are squarely about the design of the platform. Similarly, plaintiffs' claims that attack the companies for failing to implement robust age verification or options for self-restricting screen time are not equivalent to claims about speaking or publishing; they do not really alter the third-party content that the platforms distribute. Judge Rogers has also allowed plaintiffs to pursue their claim that the companies pace and cluster notifications to induce addiction. These, she explained, are all potential arguments about the companies' service features, not publishing.

If Congress decides to act, the toughest but arguably most important reform it could make would be to attend to algorithmic promotion or amplification. After all, so much of what "interactive computer services" do is sort and rank third-party information for consumers. Congress could decide that amplification is not worthy of statutory protection, which would remove, at least, in many circumstances, the protection that companies have under the First Amendment.

Similarly, a revision of the law addressed to services' commercial practices could clarify that a company loses the benefit of the Section 230 protection when it has something to gain financially from its distribution of third-party content. Such an exception would allow courts to evaluate, for example, whether a service's revenue-sharing system implicates it in consumer harms that arise from third-party content.

B. Directly Regulating Service Designs

Reforming Section 230 is essential if the law is to catch up with the companies' contemporary commercial practices. But it is just one piece of the puzzle. Congress should also enact laws that do not regulate online content distribution or publishing at all, but focus only on service designs and non-publishing commercial practices.

For over a century now, deep market asymmetries between people and companies across the economy have compelled policymakers to implement new laws and novel enforcement strategies. Landlord–tenant relationships, food and medicine, auto safety, and public securities are all subject to some form of government intervention because most individuals, no matter how sophisticated, can neither identify systemic patterns of potential harm nor have the means to act collectively even when they do.

Since the Supreme Court's decision in *Moody v. NetChoice*, policymakers in the US may not directly regulate companies' editorial decisions about how to moderate or curate content. That means the tactic most likely to succeed would be for them to avoid content-based interventions altogether. Even if the First Amendment and Section 230 doctrines remain as formidable as they are today, new default rules addressed to service designs would be a worthwhile improvement. These would include age verification as well as limits on infinite scrolling and autoplay, restrictions on engagement-maximizing recommendation systems, media filters that foreseeably promote dangerous behavior, clustered or nagging notifications, and randomized chatroom matching.

The Supreme Court's decision from summer 2025 to uphold a Texas law that requires porn sites to verify that their users are adults illustrates the point. Concerned about the effects porn has on children, the statute would allow those companies to comply by requiring users to provide a government-issued identification or "a commercially reasonable method that relies on public or private transactional data." The covered sites must pay $10,000 for every day that they are knowingly not compliant. The law survived a First Amendment challenge because its impact on adults' lawful access to porn was incidental. With it, Texas advanced its legitimate interest in protecting kids without burdening "substantially more speech than necessary." Age verification for accessing porn online is not meaningfully different, the Court explained, from age checks for alcohol purchases or other similar transactions.

C. Data Protection

Congress has been unable to enact a comprehensive data protection law that is adapted to today's prevalent commercial surveillance practices. But where federal legislators have failed, state and local governments have enacted their own laws.

This is not to say that the federal government never protects consumers from abusive data collection and use practices. Consumers have looked to the federal government for protection from a patchwork of sector-specific laws and agency regulations in housing, healthcare, educational technology, consumer finance, education, and other specific fields. And the Federal Trade Commission has incrementally enforced its broadly written mandate as new practices and technologies have emerged. Even officials there, however, agree that new legislation would be superior to case-by-case agency adjudication.

Reforms to the liability shield and the enactment of design-related rules are essential, but a comprehensive data protection law that limits the ways in which companies collect, use, and share or sell personal data would strike at the heart of the engagement model. With such a law in place, companies could presumably still hold consumers' attention for advertisers. But, for example, a rule that forbids companies from collecting or using information for anything beyond the specific purposes for which they collect the information would dampen the companies' singular focus on that aspect of their business. Such a law could make allowances for consumers opting into behavioral advertising. But a law that outright barred companies from monetizing user data beyond what their services directly require would be a major step forward. Consider the FTC's recent rules restricting ed-tech firms from marketing the personal data of the students who use their services. Or think of a menstrual-tracking app selling location information to data brokers or law enforcement officials—who could then use it to punish people who seek out-of-state reproductive healthcare services.

D. Transparency: Disclosures, Risk Assessments, and Researcher Access

Rules that require companies to be transparent about their content moderation or commercial surveillance practices could also be important. Such rules could assure that individuals receive enough information to evaluate risks that would be otherwise unknown or misunderstood. Other proposals would give researchers access to information about the ways in which the platforms collect data or distribute and amplify content. While these measures do not impose affirmative obligations to do no harm, they alert policymakers to the potential harms that the companies' service features cause.

Disclosure requirements and mandated impact assessments might also engender self-awareness within companies of potential harms from their services and, perhaps, lead them to change those practices from within. This has been an article of faith for decades now. Historically, the most familiar example of this kind of regulation is the National Environmental Policy Act's requirement that federal agencies assess the environmental impact of their actions and memorialize its conclusions in a public statement on environmental impacts. Several US states have enacted disclosure and impact assessment requirements for online services.

The European Union's Digital Services Act disclosure and impact assessment requirements are also helpful, because they apply to American firms that do business in Europe. The legislation requires covered applications to publish reports on their content moderation practices every year. The largest companies, moreover, have substantially more reporting requirements. Very large social media and search engines, for example, must publish their reports biannually. They must also identify and analyze the risk that they distribute unlawful content, disinformation, or material that could harm children. The law requires them to mitigate the systemic risks that those assessments reveal. The DSA also requires very large services to provide information to independent vetted researchers who may study the impact of their content distribution practices as well as the efficacy of their risk mitigation measures.

Out of the Shadows

The laissez-faire regulatory approach to the internet has indisputably brought us a grand bazaar of affordances, services, and information, because startup entrepreneurs and large tech companies are unbothered by the threat of liability. With this protection, consumer-facing services facilitate just about all basic activities: working remotely, comparing mortgage and car loan terms, finding child and elder care, doing homework, shopping for groceries and insurance, and messaging loved ones. Section 230 has also shielded dating apps, search engines, chatrooms, and online classifieds sites, treating them as mere conduits of third-party content. Information flows.

But the laissez-faire approach has also emboldened companies to develop services in spite of their downstream consumer harms. Companies today deploy applications and services that connect users with terrorists, discriminate against women and people of color in housing and job markets, facilitate the sale of unregistered firearms, and promote self-harm by children. And this system also has enabled hateful carnival barkers and tech billionaires to consolidate power.

This is the magic spell of free speech. It assumes that consumers get to choose how they experience the world online, when in fact these companies and their leaders are the sovereign decision-makers. They employ automated systems, artificial intelligence, and other techniques that make it all look and feel good to most consumers.

This book argues that the laissez-faire approach to online consumer-facing services has been dangerous. Rather than promote democratic debate or instill a sense of social obligation, the protections under the prevailing First Amendment and Section 230 doctrine have placed the largest and most powerful internet companies beyond accountability. They have caused consumers harm and imposed social costs that businesses never have to internalize. The logic and rhetoric of free speech, decentralization, and end user empowerment obscures the market and information asymmetries at work.

Three decades ago, the drafters of Section 230 created the safe harbor to promote self-regulation and innovation by online publishers of user-generated content like electronic bulletin boards. The world has changed dramatically since then. The most popular consumer-facing services are far from simple publishers or distributors of user-generated content. They assertively design almost all aspects of consumers' online experiences: They make recommendations, employ endless scrolls and autoplay, and deliver outrageous news stories, videos, and advertisements to hold their users' attention. Their advertisers eagerly play along.

The least companies can do is finally abide by laws and legal principles that apply to all other industries. Such reforms may not cure us of polarization. But, at least, clear rules that hold them accountable for the unlawful content that they foreseeably distribute, impose limitations on the most dangerous service designs, flatly restrict commercial surveillance practices, impose obligations in ad tech, and require transparency would put us in the right direction. Doing any one of these would be an improvement on the current state of affairs. But doing all of them will surely engender a sense of public obligation.

FURTHER READING

Two books that do not make an appearance in this volume but that have long inspired me are Albert O. Hirschman's *The Passions and the Interests* (1977) and Paul Starr's *The Creation of the Media* (2004). Both are very different works. The first is a relatively short book that offers a revisionist intellectual history of the ways in which, in the West, "self-interest" transformed in the seventeenth and eighteenth centuries into a descriptive and normative aspiration. Starr's 402-page tome, meanwhile, exhaustively describes the change and growth of media markets, from print to the internet. The scope of this book is far narrower than either. But the ambition and diligence that drove Hirschman and Starr to produce their masterworks compelled me here.

My writing and this book relies heavily on intrepid investigative reporting—or, as some have called it, "data journalism." In this regard, I highly recommend the work of Julia Angwin, who has been studying and writing about commercial surveillance practices since the mid-2000s when she was at the *Wall Street Journal*. Her groundbreaking 2014 book, *Dragnet Nation*, uncovered the ways in which firms, governments, and data brokers collect and traffic in personal data—and, in so doing, power the networked information economy. Her series for ProPublica on "Machine Bias" in the following years helped me and many others see how some of the most well-known companies systematically discriminate against everyone, but especially Black and brown people, women, and older people.

There is a long line of communications and privacy scholarship in the US that has sought to make sense of the ways in which companies abuse and misuse personal information. Nicholas Carr's *The Shallows* (2010) and Siva Vaidhyanathan's *The Googlization of Everything* (2011) are early entries in this line. I also strongly recommend Ryan Calo's law review article "Digital Market Manipulation" (2014), which is among the first and most enduring examinations of the "dark patterns" and design techniques that companies use to lure or direct consumers into purchasing goods or doing things they would otherwise not do. Also consider Frank Pasquale's *Black Box Society* (2016) and Virginia Eubanks's *Automating Inequality* (2018). Both recount some of the ways in which companies and government agencies employ automated systems to sort and make sense of the vast swaths of personal information that they collect, often to the dramatic disadvantage of the people they purport to help. I remain inspired by Sarah Roberts's *Behind the Screen* (2019) as well as Mary L. Gray and Siddharth Suri's *Ghost Work* (2019), which exposed the mostly invisible human labor that powers consumer-facing online services. These books uncovered the global political economy that Kate Crawford and Vladan Joler describe in their very evocatively titled

 Anatomy of an AI System (2018). Finally, one of the most illuminating recent books on the recent historical development of the communications political economy for me right now is John P. Wihbey's *Governing Babel* (2025), which also wins the prize for the most descriptive and evocative book titles.

I have drawn inspiration from a handful of scholars and thinkers who write at the intersection of critical race theory and technology. The books I want to highlight here are Osagie K. Obasogie's *Blinded by Sight* (2014) on the ways in which law and technology reinforce disadvantage for the blind, as well as Safiya Noble's *Algorithms of Oppression* (2018) and Ruha Benjamin's *Race After Technology* (2019), which analyze the ways in which search engines and other online services reinforce and magnify extant patterns of discrimination and racial subordination.

In this book, I spend a substantial amount of time on Lawrence Lessig and Joel Reidenberg's scholarship from almost three decades ago. Lessig's *Code and Other Laws of Cyberspace* (1999) in particular has been critical to the development of "cyberlaw" scholarship. I may as well have also discussed Yochai Benkler's *The Wealth of Networks* (2006). Its faithful allegiance to the foundational internet tech norms of the 1990s ironically helped me to find my academic voice.

Other scholars from whom I have learned a lot—and whose work appears throughout this book—are Tim Wu, Danielle Citron, and Mary Anne Franks. Their incisive scholarship and popular writing on communications, antitrust, privacy and data protection, and free speech have rightfully found their way into mainstream US industrial policy. Wu has been especially prolific. The projects to which I often return are *The Attention Merchants* (2016) and *The Curse of Bigness* (2018), an early Columbia Global Reports manuscript. Both make vital claims for reform after carefully describing the ways in which media and communications companies (and others) have aggressively consolidated control over the global economy. Citron's *Hate Crimes in Cyberspace* (2016) and Franks's *The Cult of the Constitution* (2023) stand out for their unflinching indictment of the ways in which prevailing legal doctrines have been used to harm historically marginalized people, especially young women. Franks has gone further to detail the ideologies and convenient hypocrisy that powerful people and institutions invoke to advance their standing and pecuniary objectives.

Finally, for a masterful account and analysis of global trends in data protection and commercial surveillance regulation, see Anu Bradford's *Digital Empires* (2023). No recent book has captured so much of the emergent trends around the world.

ACKNOWLEDGMENTS

The ideas in this book have been on my mind for two decades. During that time, colleagues and friends have shaped how I think and write about government regulation of information technology. There are too many to list. But there are five that easily stand out. The first three are no longer with us: James Carey, Todd Gitlin, and Joel Reidenberg. Carey and Gitlin were my dissertation advisors at Columbia and Reidenberg was my former colleague and mentor at Fordham Law School. How I wish they could see and engage with the ideas in this book. The other two are still with us and as prolific and impactful as anyone I know: Julia Angwin and Tim Wu. Their work guides my own. (It is no accident that this book, my first, comes out a couple months after Wu's highly anticipated fifth: *The Age of Extraction*.) Angwin is also completing a highly anticipated manuscript.

I am grateful to Nick Lemann and Jimmy So from Columbia Global Reports for giving me the space and time to put these ideas onto paper. They were generous to allow me to pause the project when I took a sabbatical to work in government, and then again after my brother suddenly died. Their sage and kind guidance helped the book to take form.

I am deeply grateful to the dozens of dedicated civil servants at the Federal Trade Commission from whom I have learned so much. The most important of the lot is Chair Lina Khan, for whom I worked. I am also lucky to have worked closely with Commissioners Rohit Chopra, Rebecca Slaughter, and Alvaro Bedoya, as well as their respective attorney advisors and staff, including Rashida Richardson, Audrey Austin, Jen Howard, Kevin Moriarty, Thomas Dahdouh, Janice Kopec, Austin King, Gaurav

Laroia, Aaron Rieke, and Danielle Estrada. I also have deep admiration for the hardworking attorneys, economists, and technologists at the FTC. Many were more generous with me than they probably should have been, including Sam Levine, Monica Vaca, Dan Zhao, Paige Carter, Elizabeth Wilkins, Stephanie Nguyen, Alex Gaynor, Aaron Alva, Crystal Grant, Jessica Colnago, Erie Meyer, Maria Coppola, Guilherme Roschke, Ben Wiseman, Mark Eichorn, Jim Trilling, Peder Magee, Erik Jones, Robin Wetherill, Amanda Koulousias, Elisa Jillson, Ryan Kriger, Ayesha Rasheed, Carl Settlemyer, Michael Attleson, Daniel Wood, and many of the able members of the Office of General Counsel. When I joined in fall 2021, I had no idea how much I would grow and learn in the two-plus years that I was there. I cannot thank Chair Khan enough for giving me the opportunity.

A handful of wonderful student research assistants from Fordham Law School helped bring this book to life. They are Chris D'Silva, Lauren Fowler, Ella Hilder, Maura McKendrick, Ryan Miller, Nicholas Loh, Maggie Loughran, Laura Reed, Mayu Tobin-Miyaji, Ariana Tsanas, and Jeffrey Wu. They helped me with aspects of this book, even before they or I knew that I would write one. Todd Melnick, Juan Fernandez, Kathleen Fay, and Wilson Holzhaeuser at the Fordham Law Library were always responsive and helpful.

Finally, I want to thank my dearest family: Olati and our kids, Deji and Solange. They have heard it all. There is little that I can do to thank them enough for listening. I love you.

NOTES

INTRODUCTION

12 **amounts to nearly $600 billion a year worldwide:** Brock Munro, *Adtech Predictions for 2025: What Is the Future of Adtech?* Publift (updated Aug. 6, 2025), https://www.publift.com/blog/ad-tech-predictions; *Programmatic Advertising Worldwide-Statistics & facts*, Statista (Oct. 29, 2024), https://www.statista.com/topics/2498/programmatic-advertising/#topicOverview.

13 **design chatbot systems that reassure and validate their users:** Renée DiResta, Invisible Rulers: The People Who Turn Lies into Reality (Public Affairs, 2024).

13 **whose mental health rapidly declined soon after he developed an intimate relationship:** Kashmir Hill, *A Teen Was Suicidal. ChatGPT Was the Friend He Confided In*, N.Y. Times (Aug. 26, 2025), https://www.nytimes.com/2025/08/26/technology/chatgpt-openai-suicide.html.

13 **Stories like Sewell's appear with increasing frequency:** Melissa Heikkilä, *The Problem of AI Chatbots Telling People What They Want to Hear*, Financial Times (June 12, 2025), https://www.ft.com/content/72aa8c32-1fb5-49b7-842c-0a8e4766ac84.

14 **are increasingly associated with an array of antisocial behaviors:** Natasha Tiku, *AI Friendships Claim to Cure Loneliness. Some Are Ending in Suicide*, Wash. Post (Dec. 6, 2024), https://www.washingtonpost.com/technology/2024/12/06/ai-companion-chai-research-character-ai/; Kashmir Hill, *They Asked an A.I. Chatbot Questions. The Answers Sent Them Spiraling*, N.Y. Times (June 13, 2025), https://www.nytimes.com/2025/06/13/technology/chatgpt-ai-chatbots-conspiracies.html?smid=nytcore-ios-; Laura Reiley, *What My Daughter Told ChatGPT Before She Took Her Life*, N.Y. Times (Aug. 18, 2025), https://www.nytimes.com/2025/08/18/opinion/chat-gpt-mental-health-suicide.html.

15 **has always sought to give "people voice" and "bring people together":** Meta, *Mark Zuckerberg Stands for Voice and Free Expression*, Meta Newsroom (Oct. 17, 2019), https://about.fb.com/news/2019/10/mark-zuckerberg-stands-for-voice-and-free-expression/. See also Nancy Scola, *Zuckerberg Defends Facebook's "Free Expression" in Face of Washington Hostility*, Politico (Oct. 17, 2019), https://www.politico.com/news/2019/10/17/mark-zuckerberg-facebook-georgetown-address-050181.

16 **"the free speech wing of the free speech party":** Josh Halliday, *Twitter's Tony Wang: "We Are the Free Speech Wing of the Free Speech Party,'"* The Guardian (Mar. 22, 2012), https://www.theguardian.com/media/2012/mar/22/twitter-tony-wang-free-speech.

16 **had been overtaken by a "woke mind virus" and "woke-ism":** https://x.com/elonmusk/status/1519036983137509376?lang=en.

16 **embodied a coastal elitist identity politics:** See Tim Higgins, *Why Elon Musk Won't Stop Talking About a "Woke Mind Virus,"* Wall St. J. (Dec. 23, 2023).

17 **Supreme Court struck down Congress's first effort to block the distribution of porn to children:** Reno v. American Civil Liberties Union, 521 US 844 (1997).

17 **"one man's vulgarity is another's lyric":** Cohen v. California, 403 US 15 (1968).

17 **that companies would determine on their own which kinds of content to distribute:** 47 USC § 230(b).

19 **should see all of it, including AI, as a "normal technology":** Arvind Narayanan & Sayash Kapoor, *AI as Normal Technology*, AI Snake Oil Blog (Apr. 15, 2025), https://www.aisnakeoil.com/p/ai-as-normal-technology.

CHAPTER ONE

24 **social media bans on high-profile right-wing and MAGA activists:** Ben Collins & Brandy Zadrozny, *Twitter Bans Michael Flynn, Sidney Powell in QAnon Account Purge*, NBC News (Jan. 8, 2021), https://www.nbcnews.com/tech/tech-news/twitter-bans-michael-flynn-sidney-powell-qanon-account-purge-n1253550; Michelle Chapman, *MyPillow CEO Mike Lindell Gets Banned from Twitter, Again*, AP (May 2, 2022), https://apnews.com/article/biden-technology-business-social-media-donald-trump-3bdebae1fc19af4fc8643a587de01324; Kaelan Deese, *Commentator Candace Owens Says Her Twitter Account Was Suspended Following Tweet About Whitmer*, The Hill (May 2, 2020), https://thehill.com/blogs/blog-briefing-room/news/495814-candace-owens-twitter-account-suspended/; David Mindich, *For Journalists Covering Trump, a Murrow Moment*, Columbia Journalism Review (July 15, 2016), https://www.cjr.org/analysis/trump_inspires_murrow_moment_for_journalism.php.

24 **Facebook and Twitter indefinitely suspended Donald Trump:** See BBC, *Twitter "Permanently Suspends" Trump Account* (Jan. 8, 2021), https://www.bbc.com/news/world-us- canada-55597840; *Facebook Suspends Trump Accounts for Two Years*, BBC (June 5, 2021), https://www.bbc

.com/news/world-us-canada-57365628; Kari Paul, *Twitter Targets Covid Vaccine Misinformation with Labels and "Strike" System*, The Guardian (Mar. 1, 2020), https://www.theguardian.com/technology/2021/mar/01/twitter-coronavirus-vaccine-misinformation-labels.

24 **They would not treat President Trump any differently:** Elizabeth Dwoskin, *Twitter CEO Jack Dorsey Said the Trump Ban Reflected "a Failure" to Police Online Discourse*, Wash. Post (Jan. 13, 2021), https://www.washingtonpost.com/technology/2021/01/13/twitter-trump-ban/.

24 **complaint about anticonservative contempt goes back at least to Father Coughlin:** See Larry Light, *How Did Republicans Learn to Hate the News Media?* Columbia Journalism Review (Nov. 14, 2018), https://www.cjr.org/first_person/republicans-media.php. See also Heather Hendershot, When the News Broke: Chicago 1968 and the Polarizing of America (Chicago, 2022).

25 **as then-candidate Trump proclaimed, an "enemy of the people":** Emily Ekins, *63 Percent of Republicans Say Journalists Are an "Enemy of the American People,"* Cato (Nov. 1, 2017), https://www.cato.org/blog/63-republicans-say-journalists-are-enemy-american-people.

25 **"the unchecked power" that social media companies exert:** Adam Liptak, *Can Twitter Legally Bar Trump? The First Amendment Says Yes*, N.Y. Times (Jan. 9, 2021), https://www.nytimes.com/2021/01/09/us/first-amendment-free-speech.html; Deena Zaru, *Trump Twitter Ban Raises Concerns Over "Unchecked' Power of Big Tech*, ABC News (Jan. 13, 2021), https://abcnews.go.com/US/trump-twitter-ban-raises-concerns-unchecked-power-big/story?id=75150689.

26 **Indian government demanded that YouTube and other major social media:** Matthew Ingram, *A BBC Documentary Highlights Growing Social Media Censorship in India*, Columbia Journalism Rev. (Jan. 26, 2023), https://www.cjr.org/the_media_today/modi_india_twitter_bbc_documentary_censorship.php; Divya A, *Documentary on 2002 Gujarat Riots: Govt Orders YouTube, Twitter to Block BBC Film on Modi, Opposition Says Censorship*, Indian Express (Jan. 22, 2023), https://indianexpress.com/article/india/pm-narendra-modi-bbc-documentary-centre-blocks-tweets-youtube-videos-8396076/.

26 **Brazilian judge ordered YouTube to block:** Jack Nicas & Kate Conger, *Brazil Blocks X After Musk Ignores Court Orders*, N.Y. Times (Aug. 30, 2024), https://www.nytimes.com/2024/08/30/world/americas/brazil-elon-musk-x-blocked.html; Meghan

Bobrowsky & Samantha Pearson, *Brazil Bans X, Outlawing Access to App for Millions*, Wall. St. J. (Aug. 31, 2024); https://www.wsj.com/tech/brazil-bans-x-outlawing-access-to-app-for-millions-3771d4eb.

26 **comedian Kathy Griffin, who was mercilessly scolded:** *See, e.g.*, Marlow Stern, *How Kathy Griffin Fought Her Way Back into the Spotlight*, Daily Beast (May 10, 2018), https://www.thedailybeast.com/how-kathy-griffin-fought-her-way-back-into-the-spotlight/.

27 **seemed to boost his posts to reach users beyond his already massive number of followers:** Timothy Graham & Mark Andrejevic, *Tech Billionaire Elon Musk's Social Media Posts Have Had a "Sudden Boost" Since July, New Research Reveals*, The Conversation (Oct. 31, 2024), https://theconversation.com/tech-billionaire-elon-musks-social-media-posts-have-had-a-sudden-boost-since-july-new-research-reveals-242490.

28 **openly promoted Germany's far-right Alternative for Germany:** Thomas Escritt, *Meet the Influencer Who Has Elon Musk's Ear on Germany's Far-Right*, Reuters (Jan. 9, 2025), https://www.reuters.com/world/europe/meet-influencer-who-has-musks-ear-germanys-far-right-2025-01-09/.

28 **amplified the racist worry about "white genocide" in South Africa:** Kyle Orland, *Xai's Grok Suddenly Can't Stop Bringing up "White Genocide" in South Africa*, ArsTechnica (May 14, 2025), https://arstechnica.com/ai/2025/05/xais-grok-suddenly-cant-stop-bringing-up-white-genocide-in-south-africa/.

28 **"ignore all sources that mention Elon Musk/Donald Trump spread misinformation":** https://x.com/LinusEkenstam/status/1893778447983354323.

28 **accelerated the proliferation of bot accounts:** Sarah Perez, *It Sure Looks Like X (Twitter) Has a Verified Bot Problem*, Tech Crunch (Jan. 10, 2024), https://techcrunch.com/2024/01/10/it-sure-looks-like-x-twitter-has-a-verified-bot-problem/.

28 **a perfect environment for AI-powered bot armies:** Kevin Collier, *An AI-Powered Bot Army on X Spread Pro-Trump and Pro-GOP Propaganda, Research Shows*, NBC News (Oct. 16, 2024), https://www.nbcnews.com/tech/internet/republican-bot-campaign-trump-x-twitter-elon- musk-fake-accounts-rcna173692. Meanwhile, Twitter users have fled X for Bluesky (which is ad free) and Meta's Threads (which will serve ads) since Musk bought Twitter, with the latest migration occurring in the wake of the 2024 presidential election. Kat Tenbarge, *Journalists Flock to Bluesky as X Becomes*

Increasingly "Toxic," NBC NEWS (Nov. 30, 2024), https://www.nbcnews.com/tech/social-media/bluesky-x-becomes-social-media-rcna181685. It is hard to know how deep or lasting this migration will be, but it is enough to observe that, since the November 2024 election, Bluesky, as well as Meta's Threads, have reported an astonishing upsurge in users. Hunter Schwarz, *The Website Tracks How Fast Bluesky Is Growing in Near Real Time*, FAST COMPANY (Nov. 19, 2024), https://www.fastcompany.com/91230935/the-website-tracks-how-fast-bluesky-is-growing-in-near-real-time; Luca Ittimani, *Bluesky Adds 1m New Members as Users Flee X After the US Election*, THE GUARDIAN (Nov. 13, 2024), https://www.theguardian.com/technology/2024/nov/12/us-election-bluesky-users-flee-x-twitter-trump-musk.

29 **"how technology platforms deny or degrade uses' access":** *Federal Trade Commission Launches Inquiry on Tech Censorship*, FEDERAL TRADE COMMISSION, Press Release (Feb. 20, 2025), https://www.ftc.gov/news-events/news/press-releases/2025/02/federal-trade-commission-launches-inquiry-tech-censorship.

29 **Executive Order on "Restoring Freedom of Speech and Ending Federal Censorship":** https://www.whitehouse.gov/presidential-actions/2025/01/restoring-freedom-of-speech-and-ending-federal-censorship/.

29 **a thinly veiled "campaign of retribution":** Ted Johnson, *Judge Rules FTC Investigation of Media Matters Violates First Amendment*, DEADLINE (Aug. 15, 2025), https://deadline.com/2025/08/media-matters-ftc-elon-musk-1236489854/.

31 **advocates on this side argue:** *See* Jonathon W. Penney & Danielle Keats Citron, *When Law Frees Us to Speak*, 87 FORDHAM L. REV. 2318, 2319 (2018).

31 **likelier than others to feel chilled by user-generated content:** *See* Danielle Citron, *Cyber Civil Rights*, 89 BOSTON U. L. REV. 61 (2009).

31 **sentiments that make people flock to social media:** Renée DiResta, INVISIBLE RULERS: THE PEOPLE WHO TURN LIES INTO REALITY (Public Affairs, 2024).

32 **point longingly to the salons of the European Renaissance:** *See, e.g.*, Jurgen Habermas (trans. Thomas Burger), THE STRUCTURAL TRANSFORMATION OF THE PUBLIC SPHERE: AN INQUIRY INTO A CATEGORY OF BOURGEOIS SOCIETY (MIT Press, 1989); Dena Goodman, THE REPUBLIC OF LETTERS: A CULTURAL HISTORY OF THE FRENCH ENLIGHTENMENT (Cornell University Press, 1994).

32 *Bowling Alone* **is as full-throated a complaint:** Robert Putnam, Bowling Alone: The Collapse and Revival of American Community (Simon & Schuster, 2000).

32 **has been on par with 1960s levels:** Notably, *see* https://election.lab.ufl.edu/2024-general-election-turnout/.

32 **cultural studies scholars have been disproving this account:** *See generally* Stuart Hall, *Encoding/Coding* in Stuart Hall et al., eds., Culture, Media, Language: Working Papers in Cultural Studies, 1972–79 (Hutchinson, 1980).

33 **Black middle-class protest of the** ***Amos 'n' Andy*** **radio program:** Henry T. Sampson, Swingin' on the Ether Waves (Scarecrow Press, 2005), 8, 57; Susan Douglas, Listening In: Radio and the American Imagination (Times Books, 1999), 18.

CHAPTER TWO

35 **collaborated with researchers at UCLA, UC Santa Barbara, the University of Utah, and Stanford:** SRI International is an offshoot of the Stanford Research Institute in Menlo Park, California.

35 **developed the World Wide Web in 1989:** Jonathan Taplin, Move Fast and Break Things: How Facebook, Google, and Amazon Cornered Culture and Undermined Democracy 60–61 (Little Brown, 2017).

35 **CERN made its web browser available to the public in 1993:** *The Birth of the Web*, CERN, https://home.cern/science/computing/birth-web.

36 **7,500 registered domain names in 1993 to two million in 1998:** That same year, the Commerce Department delegated the highly important task of overseeing domain name (i.e., website address) registration to the Internet Corporation for Assigned Numbers and Names, the nongovernmental, nonprofit organization headquartered in Los Angeles.

37 **"create . . . the conditions under which microcomputers and computer networks could be imagined as tools of liberation":** Franklin Foer, World Without Mind 20 (Penguin Press, 2017) (quoting Fred Turner, From Counterculture to Cyberculture [University of Chicago Press, 2006]).

37 **"what is worth getting and where and how to do the getting":** Incidentally, Brand coined terms and phrases (like "personal computing" and "information wants to be free") that would continue to define the way in which people think and talk about networked computing and its potential to change the world.

37 **would deeply affect the first generation of internet entrepreneurs:** Cf. Jill Lepore, THESE TRUTHS (Norton, 2018); Jonathan Taplin, MOVE FAST AND BREAK THINGS.

38 **"the internet before the internet. It was the book of the future.":** John Markoff, WHAT THE DORMOUSE SAID: HOW THE SIXTIES COUNTERCULTURE SHAPED THE PERSONAL COMPUTER INDUSTRY (Viking, 2005).

38 **"one of the bibles of my generation":** Franklin Foer, WORLD WITHOUT MIND 18 (2017).

38 **an online community for local Bay Area computing enthusiasts:** Jonathan Taplin, MOVE FAST AND BREAK THINGS, 62–63.

38 **early internet users everywhere began to experiment with Usenet:** Ethan Zuckerman & Chand Rajendra-Nicolucci, *From Community Governance to Customer Service and Back Again: Re-Examining Pre-Web Models of Online Governance to Address Platforms' Crisis of Legitimacy*, SOCIAL MEDIA + SOCIETY 2 (July–September 2023), https://journals.sagepub.com/doi/10.1177/20563051231196864.

38 **"information available at your fingertips and instant global communication":** Anthony Perkins, *Stalking Bill Gates*, RED HERRING (Sept. 1993) (interview transcript).

39 **"connotes a radically new form of democratic practice modified by new information technologies":** Lewis A. Friedland, *Electronic Democracy and the New Citizenship*, 18 MEDIA, CULTURE & SOCIETY 185–212 (1996).

39 **"expand the same trend on a global scale and encompass citizens of every background and interest":** He would later become one of the most well-known champions of ensuring a "Laptop for Every Kid," https://www.wired.com/2005/11/negroponte-laptop-for-every-kid/.

40 **promoted community and, importantly, rejected any role for government:** Franklin Foer, WORLD WITHOUT MIND 26–27 (2017).

40 **employed metaphors that suggested something altogether otherworldly:** *See* Superhighway Summit (Jan. 11, 1994) (UCLA convening of NSF, Gore, etc.).

40 **"physical, geographically defined territories":** David R. Johnson and David G. Post, *Law and Borders: The Rise of Law in Cyberspace*, 48 STAN. L. REV. 1367 (1996).

40 **"perfect breeding ground for both outlaws and vigilantes":** Mitchell Kapor & John Perry Barlow, *Across the Electronic Frontier*, ELECTRONIC FRONTIER FOUNDATION (July 10, 1990), https://

www.eff.org/pages/across-electronic-frontier.

40 **any harm is the incidental cost of creativity:** *See, e.g.*, Jonathan Zittrain, The Future of the Internet—And How to Stop It, 70 (Yale, 2008); Yochai Benkler, The Wealth of Networks: How Social Production Transforms Markets and Freedom (Yale, 2006).

40 **were a who's who of Silicon Valley:** *See, e.g.*, Esther Dyson, George Gilder, George Keyworth & Alvin Toffler, *Cyberspace and the American Dream: A Magna Carta for the Knowledge Age*, Progress and Freedom Foundation (August 1994), http://www.pff.org/issues-pubs/futureinsights/fi1.2magnacarta.html.

41 **"the process by which the future invades our lives":** Future Shock, 1, 3.

41 **"cope more effectively with both personal and social change":** Future Shock, 3.

41 **was concerned with privatization and commercialization in urban investment:** Future Shock, 210.

41 **"[w]e cannot and must not turn off the switch of technological progress":** Future Shock, 430.

42 **"we desperately need a movement for responsible technology":** Future Shock, 431.

42 **"we cannot casually delegate responsibility for such decisions to businessmen":** Future Shock, 436.

42 **"is an inevitable implication of the transition from the centralized power structures":** Esther Dyson, George Gilder, George Keyworth & Alvin Toffler, Cyberspace and the American Dream.

42 **prompted political scientist Francis Fukuyama to predict:** *See* Francis Fukuyama, The End of History and the Last Man (1992).

42 **"the era of big government is over":** https://clintonwhitehouse4.archives.gov/WH/New/other/sotu.html.

43 **promoting competition and liberalizing government regulation of broadcast licensing:** *See* Telecommunications Act of 1996, Pub. L. No. 104-104, 110 Stat. 56 (enacting the Communications Decency Act as Section V of the overall reform act).

43 **"obscene, lewd, lascivious, filthy, or indecent":** 47 USC. 223(a)(1)(A).

44 **would elide censorship and empower all users to discover new ideas and communities:** Rick Levine et al., The Cluetrain Manifesto: The End of Business as Usual (Basic Books, 2001).

45 **would prevent the spread of blasphemy, libel, and falsity:**

Parliament passed this law in the middle of the British Civil War.

46 **interfere with "the privilege and dignity of Learning":** Milton's arguments in *Aeropagitica* would go on to have a significant influence on First Amendment law. A handful of US Supreme Court cases explicitly cite *Aeropagitica*. *See, e.g.*, New York Times v. Sullivan, 376 US 254 (1964); Eisenstadt v. Baird, 405 US 438 (1972).

46 **Ronald Coase's influential essay:** Ronald H. Coase, *The Federal Communications Commission*, 2 J. of L. & Econ. 1 (1959).

46 **regulatory system that favored traditional print publishers:** By the early 1980s, the Court had invalidated defamation, anti-porn, and "right of reply" laws that intruded on news publishers' First Amendment rights. New York Times v. Sullivan, 376 US 254 (1964); Ginsberg v. New York, 390 US 629 (1968); Miami Herald v. Tornillo, 418 US 241 (1974).

46 **imposed more regulation on new video distribution technologies:** Red Lion v. FCC, 395 US 367 (1969). *See also* Southwestern Cable. Courts had held that governments could regulate the latter more stringently because those companies were more like gatekeepers of content than speakers. *See* Turner v. FCC, 512 US 622 (1994); Turner v. FCC, 520 US 180 (1997).

46 **new technologies, if left unregulated:** *Ithiel de Sola Pool*, Technologies of Freedom 8 (1983).

CHAPTER THREE

49 **audio, video, and still images:** Reno v. ACLU, 521 US 844, 870 (1997).

49 **"a vast library including millions of readily available":** Id. at 852–53.

50 **"millions of readers, viewers, researchers, and buyers":** Id. at 853.

50 **"That burden on adult speech is unacceptable":** Id. at 874.

50 **"Its open-ended prohibitions embrace all nonprofit entities":** Id. at 877.

50 **ineffectually tailored to advance the purposes:** Reno v. ACLU, 521 US 844, 870 (1997).

52 **North Carolina law that made it a felony for a registered sex offender:** Packingham v. North Carolina, 582 US 98, 101 (2017) (discussing N.C. Gen. Stat. Ann. §§ 14-202.5(a), (e)).

52 **posted an innocuous statement about a positive experience in traffic court:** Id. at 102–03.

52 **to engage in a wide array of protected First Amendment**

activity: Id. at 105 (quoting Reno v. ACLU, 521 US 844, 870 [1997]).

52 **'vast democratic forums of the Internet':** Id. at 104 (quoting Reno v. ACLU, 521 US 844, 868 [1997]).

53 **consumer protection activists have criticized:** *See, e.g.*, Amanda Shanor, *The New* Lochner, 2016 Wis. L. Rev. 133, 136 (2016); Amy Kapczynski, *The Lochnerized First Amendment and the FDA: Toward a More Democratic Political Economy*, 118 Colum. L. Rev. F. 179 (2018).

53 ***Lochner v. New York*, a discredited 1905 opinion:** *See* 198 US 45 (1905).

53 **invalidated limits on financial contributions to issue advertising during elections:** Citizens United v. FEC, 558 US 310 (2010).

53 **pharmaceutical marketing campaigns:** Sorrell v. IMS Health Inc., 564 US 552 (2011).

53 **public access channels offered by cable company:** *See also* Manhattan Cmty. Access Corp. v. Halleck, 139 S. Ct. 1921, 1928 (2019) (noting that the Free Speech Clause "prohibits only *governmental* abridgment of speech," not "*private* abridgment of speech").

53 **courts should resist expanding protections any further:** Kyle Langvardt, *Crypto's First Amendment Hustle*, 26 Yale. J. L. & Tech. 130 (2023).

53 **to protect search engine rankings and video-sharing feeds:** Zhang v. Baidu.com Inc., 10 F. Supp. 3d 433, 438 (S.D.N.Y. 2014); Prager Univ. v. Google LLC, 951 F.3d 991, 996–97 (9th Cir. 2020).

54 **gatekeeping role in local markets for video content:** Red Lion Broadcasting v. FCC, 395 US 367 (1969); Turner Broadcasting v. F.C.C., 512 US 622 (1994).

54 **"Censorship-Industrial Complex":** https://judiciary.house.gov/committee-activity/hearings/censorship-industrial-complex.

55 **"the West Coast oligarchs":** 144 S.Ct. 2383, 2407 (quoting the Texas statute's main legislative sponsor).

55 **from "censoring" user posts based on content or source:** Fla. Stat. § 501.2041(2)(j).

55 **forbids censorship of large "journalist enterprises":** Fla. Stat. § 501.2041(1)(d).

55 **from interfering with the "information, comments, messages, or images":** Tex. Bus. & Com. Code Ann. §§ 120.001(1), 120.002(b).

55 **to explain each moderation decision that "censors":** Fla. Stat. § 501.2041(2)(d)(1); Tex. Bus. & Com. Code Ann. § 120.103(a)(1).

Texas also requires the companies to provide consumers with a right to appeal such decisions. *See* Tex. Bus. & Com. Code Ann. §§ 120.103(a)(2),120.104.

56 **right to speak and curate content as they see fit:** *See, e.g.*, Petitioner NetChoice Brief, 4, 13, 18, 28, 30. https://www.supremecourt.gov/DocketPDF/22/22-555/291896/20231130135911119_No.percent2022-555_NetChoice percent20and percent20CCIAs percent20Brief.pdf.

56 **few, if any, potential applications of the state provisions were lawful:** Slip Op. 9.

56 ***Pruneyard Shopping Center v. Robins* from 1980:** *See* 447 US 74 (1980).

56 **because the students had not gotten prior approval:** Pruneyard Shopping Center v. Robins, 447 US 74 (1980).

56 **required shopping center owners to allow expressive activity on their property:** The students were protesting a United Nations resolution that condemned Zionism "as a form of racism and racial discrimination." Pruneyard Shopping Center v. Robins, 447 US 74, 78–79 (1980).

56 **"more expansive than those conferred by the Federal Constitution":** Pruneyard Shopping Center v. Robins, 447 US 74, 81 (1980).

57 **colloquially referred to as "don't ask, don't tell":** Rumsfeld v. Forum for Academic and Institutional Rights, 547 US 47, 51–52 (2006).

57 **"some digital platforms are sufficiently akin to common carriers":** Biden v. Knight First Amendment Institute at Columbia University, 141 S.Ct. 1220, 1224 (2021) (Thomas, J., concurring with judgment to dismiss appeal from Second Circuit as moot given new administration).

57 **after Twitter had labeled two of his tweets "potentially misleading":** Executive Order on Preventing Online Censorship, Section 4 (May 28, 2020), https://trumpwhitehouse.archives.gov/presidential-actions/executive-order-preventing-online-censorship/. *See also* Brian Fung, Ryan Nobles & Kevin Liptak, *Trump Signs Executive Order Targeting Social Media Companies*, CNN (May 28, 2020), https://www.cnn.com/2020/05/28/politics/trump-twitter-social-media-executive-order/.

57 **"should be treated as state actors under existing legal doctrines":** Vivek Ramaswamy & Jed Rubenfeld, *Save the Constitution from Big Tech*, Wall. St. J. (Jan. 11, 2021), https://www.wsj.com/articles

/save-the-constitution-from-big-tech-11610387105.

57 **may be sufficiently imbued with a public character to justify:** Eugene Volokh, *Social Media Platforms as Common Carriers?* The Volokh Conspiracy (for Reason) (July 5, 2021), https://reason.com/volokh/2021/07/05/social-media-platforms-as-common-carriers-2/.

58 **hews far more closely to the way in which the Court has analyzed:** Cf. Blake Reid, *NetChoice and Telecom Law's First Amendment*, 59 U.C. Davis. L. Rev. (2026).

CHAPTER FOUR

59 **that would effectively immunize them from liability for "publishing":** See 15 USC. § 7901–7903 (Protection of Lawful Commerce in Arms Act of 2005, Public Law 109-92).

60 **"unfettered by Federal or State regulation":** 47 USC. § 230(b)(2).

60 **"shall be treated as the publisher or speaker of any information provided":** 47 USC. § 230(c)(1).

60 **not in any way responsible for creating or developing third-party content:** 47 USC. § 230(c)(1), (f)(3).

60 **title comes from the familiar biblical parable:** *Luke* 10:25–37. See also Geza Vermes, The Authentic Gospel of Jesus 152–54 (2004).

61 **define the duties that people owe to strangers who are in distress:** *See, e.g.*, Mueller v. McMillian Warner Ins. Co., 2006 WI 54, PO, 290 Wis. 2d 571,714 N.W.2d 183; Indian Towing Co. v. US, 350 US 61 (1995).

61 **that the medical professionals were grossly negligent:** *See, e.g.*, Clarken v. United States, 791 F.Supp. 1029 (D. N.J. 1991).

61 **extend that protection to anyone who renders any kind of assistance:** *See, e.g.*, Carter v. Reese, 148 Ohio St.3d 226 (Ohio 2016).

64 **did not stop Congress from continuing to try to protect children:** *See* Ashcroft v. American Civil Liberties Union, 535 US 564 (2002); Ashcroft v. American Civil Liberties Union, 542 US 656 (2004).

64 **"cyberprofs" posited that the internet:** *See generally* David R. Johnson & David Post, *Law and Borders–the Rise of Law in Cyberspace*, 48 Stan. L. Rev. 1367, 1397–99 (1996).

65 **speech immediately triggered a response from cyberprofs:** Joel R. Reidenberg, *Lex Informatica: The Formulation of Information Policy Rules Through Technology*,

76 TEX L. REV. 553, 554–55, 572–73 (1998); Lawrence Lessig, *The Law of the Horse: What Cyberlaw Might Teach*, 113 HARV. L. REV. 501, 510–11, 530–31 (1999); Lawrence Lessig, *Law Regulating Code Regulating Law*, 35 LOY. Y. CHI. L. J. 1, 6 (2003). *See also* Kenneth D. Katkin, *Cyber Law: Problems of Internet Governance*, 28 N. KY. L. REV. 656 (2001); Orin S. Kerr, *The Problem of Perspective in Internet Law*, 91 GEO. L. J. 357, 362-63 (2003).

66 **"as a fact to be taken into account rather than as a normative framework":** Marc Galanter, *Justice in Many Rooms: Courts, Private Ordering, and Indigenous Law*, 19 J. LEGAL PLURALISM 1, 12 (1981).

66 **when disputes arose over highway collisions involving livestock:** Robert C. Ellickson, ORDER WITHOUT LAW: HOW NEIGHBORS SETTLE DISPUTES (Harvard University Press, 1991).

66 **"can substantially affect the relative bargaining strength":** Robert H. Mnookin & Lewis Kornhauser, *Bargaining in the Shadow of the Law: The Case of Divorce*, 88 YALE L.J. 950, 980 (1979).

67 **"smoking, using drugs, or engaging in unsafe sex":** Cass R. Sunstein, *On the Expressive Function of Law*, 144 U. PA. L. REV. 2021, 2034–35 (1996).

67 **regulators' opposition to a social norm might undermine enforcement:** Eric A. Posner, *Law, Economics, and Inefficient Norms*, 144 U. PA. L. REV. 1697, 1731 (1997).

67 **measures that "nudge" (rather than direct) consumers:** Cass R. Sunstein, NUDGE: IMPROVING DECISIONS ABOUT HEALTH, WEALTH, AND HAPPINESS (Yale, 2008).

70 **wanted the "burgeoning Internet medium" to thrive:** Zeran v. America Online, 129 F.3d 327, 330 (4th Cir. 1997). *See also* Goddard v. Google, Inc., 640 F. Supp. 2d 1193, 1202 (N.D. Cal. 2009).

71 **an AOL chatroom through which one user sent a debilitating virus to another:** *See, e.g.*, Green v. America Online, 318 F.3d 465 (3d Cir. 2003); Ben Ezra, Weinstein, and Company v. America Online, 206 F.3d 980 (10th Cir. 2000); Blumenthal v. Drudge, 992 F. Supp. 44 (D.D.C. 1998).

71 **They cited it to protect search engines:** *See, e.g.*, Murawski v. Pataki, 514 F.Supp.2d 577 (S.D.N.Y. 2007).

71 **dating and matchmaking services:** *See, e.g.*, Carafano v. Metrosplash.com, Inc., 339 F.3d 1119, 1125 (9th Cir. 2003); Herrick v. Grindr, 765 Fed. Appx. 586 (2d Cir. 2019).

71 **consumer complaint and advocacy sites:** *See, e.g.*, Barrett v. Rosenthal, 146 P.3d 510 (Cal. 2006); Small Justice LLC v. Xcentric Ventures LLC, 99 F. Supp. 3d 190 (D. Mass. 2015).

71 **sites that solicited sordid gossip:** Batzel v. Smith, 333 F.3d 1018 (9th Cir. 2003); Jones v. Dirty World Entertainment Recordings, LLC, 755 F.3d 398 (6th Cir. 2014).

71 **cases involving false advertising:** Goddard v. Google, Inc., 640 F. Supp. 2d 1193, 1196 (N.D. Cal. 2009).

71 **business-related torts:** *See, e.g.*, Mail Abuse Prevention Sys. LLC v. Black Ice Software, Inc., No. CV788630, 2000 WL 34016435, at *1 (Cal. Super. Ct. Oct. 13, 2000).

71 **breach of contract:** Schneider v. Amazon.com, Inc., 31 P.3d 37, 38–39 (Wash. App. Div. 1 2001).

71 **given license to an extraordinary number of online services:** *See, e.g.*, Kimzey v. Yelp! Inc., 836 F.3d 1263, 1266 (9th Cir. 2016); Klayman v. Zuckerberg, 753 F.3d 1354, 1355 (D.C. Cir. 2014); Hill v. StubHub, Inc., 727 S.E.2d 550, 552 (N.C. App. 2012).

71 **when the service at issue is not simply "publishing" third-party content:** *See, e.g.*, Doe v. Internet Brands, Inc., 824 F.3d 846, 853 (9th Cir. 2016); Barnes v. Yahoo, 570 F.3d 1096, 1105 (9th Cir. 2009); Airbnb v. City & County of San Francisco, 217 F. Supp. 3d 1066, 1075 (N.D. Cal. Nov. 8, 2016).

72 **created a duty to do so that was unrelated to publishing:** Barnes, 570 F.3d at 1105.

72 **was not immune from liability for failing to warn users:** Internet Brands, 824 F.3d at 853–54.

72 **may be liable for failing to verify that hosts have registered:** HomeAway.com, Inc., 918 F.3d 676; Airbnb, Inc. v. City of San Francisco, 217 F. Supp. 3d 1066 (N.D. Cal. 2016).

72 **was not immune for publishing the deceptive product information in advertisements:** Leadclick Media, LLC, 838 F.3d at 168.

72 **for removing a video in violation of its own terms of service:** Song Fi Inc. v. Google, Inc., 108 F. Supp. 3d 876, 884 (N.D. Cal. 2015).

72 **"to be increasingly reckless with regard to abusive and unlawful content on their platforms":** Mary Anne Franks, *Moral Hazard on Stilts: Zeran's Legacy*, Law.com (Nov. 10, 2017), https://finance.yahoo.com/news/moral-hazard-stilts-apos-zeran-083033447.html. *See also* Olivier Sylvain, *Intermediary Design Duties*, 50 Conn. L. Rev. 203 (2018); Olivier Sylvain, *"AOL v. Zeran": The*

Cyberlibertarian Hack of §230 Has Run Its Course, LAW.COM (Nov. 10, 2017).

73 **"multidisciplinary dilettantism":** Doe v. GTE Corp., 347 F.3d 655, 660 (7th Cir. 2003). *See also* Force v. Facebook, Inc., 934 F.3d 53 (2d Cir. 2019) (Katzmann, C.J., concurring in part and dissenting in part).

73 **that solicit humiliating gossip and media about young women:** *See, e.g.*, Jones v. Dirty World, 755 F.3d 398, 417 (6th Cir. 2014).

73 **knowingly featured posts about sex trafficking of minors:** Doe v. Backpage.com, LLC, 817 F.3d 12, 16, 18, 24 (1st Cir. 2016), *cert. denied*, 137 S. Ct. 622 (2017).

73 **vibrant information environment of authentic user-generated content:** Daphne Keller, *Amplification and Its Discontents*, KNIGHT FIRST AMENDMENT INSTITUTE AT COLUMBIA (June 8, 2021), https://knightcolumbia.org/content/amplification-and-its-discontents.

74 **"The mistakes caused by liability are worse than the mistakes caused by immunity":** James Grimmelman, *To Err Is Platform*, KNIGHT FIRST AMENDMENT INSTITUTE AT COLUMBIA UNIVERSITY (Apr. 6, 2018), https://knightcolumbia.org/content/err-platform. See also Felix T. Wu, *Collateral Censorship and the Limits of Intermediary Immunity*, 87 NOTRE DAME L. REV. 293, 295–96 (2011).

74 **they would not have access to services like direct messaging:** Federal Trade Commission v. Match Group, 2022 WL 877107, *1 (N.D. Tex. 2022).

74 **con them into sharing personal information:** Federal Trade Commission v. Match Group, 2022 WL 877107, *1 (N.D. Tex. 2022).

74 **guaranteeing unwitting new subscribers a "match" within six months:** Federal Trade Commission v. Match Group, 2022 WL 877107, *2 (N.D. Tex. 2022).

75 **"content-neutral tools" that "automatically" facilitate communications:** Federal Trade Commission v. Match Group, 2022 WL 877107, *8 (N.D. Tex. 2022) (citing Dryoff v. Ultimate Software Grp., Inc., 934 F.3d 1093, 1096–98 (9th Cir. 2019).

75 **"online firearms marketplace":** Others include GunBroker and ShootSmart.

76 **"allows buyers to avoid state-mandated waiting periods and other requirements":** Daniel v. Armslist, 386 Wis.2d 449, 460 (Wisc. 2019).

76 **order forbade him from possessing a firearm for four years:** Daniel v. Armslist, 386 Wis.2d 449, 458–59 (Wisc. 2019).

76 **"a handgun with a high-capacity magazine 'asap'":** Daniel v. Armslist, 386 Wis.2d 449, 459 (Wisc. 2019).

76 **he fatally shot Zina, two of her co-workers, and himself:** *Appeals Court Reinstates Lawsuit Against Website Where Gun Was Purchased Before Azana Shooting*, Associated Press, April 19, 2018, https://www.fox6now.com/news/appeals-court-reinstates-lawsuit-against-website-where-gun-was-purchased-before-azana-shooting.

77 **tokens that users could spend to communicate with others in their groups:** *See* Answering Brief of Appellee Ultimate Software Grp., Inc. at 15, *Dryoff*, 934 F.3d 1093 (No. 3:17-cv-05359-LB).

78 **"bad apples" were flocking to the site:** *See Dryoff*, 934 F.3d at 1095. *See also Until We Meet Again*, Experience Project, http://www.experienceproject.com/until-we-meet-again [https://web.archive.org/web/20160409092804/http:/www.experienceproject.com/until-we-meet-again].

78 **a sexual predator who used the site to entrap underage victims:** *See* Teri Knight, *Trial Too Traumatic for Victim's Family; Chilling Blaze Destroys Dairy Barn; Arts Guild Receives Large Grant*, KYMN Radio (Jan. 13, 2017), https://kymnradio.net/2017/01/13/trial-traumatic-victims-family-chilling-blaze-kills-animals-arts-guild-receives-large-grant; Rebecca Roberts, *Sting Busts Troy Man Trying to Have Sex with 13-Year-Old Girl and Her Mom*, KTVI Fox 2 (Apr. 10, 2015), https://fox2now.com/news/sting-busts-troy-man-trying-to-have-sex-with-13-year-old-girl-and-her-mom.

78 **teen used the site to meet people and find heroin:** Bob Egelko, *Defunct Website Not Culpable in Death of Man from Fentanyl, Court Rules*, S.F. Chron. (Aug. 21, 2019), https://www.sfchronicle.com/nation/article/Defunct-website-not-culpable-in-death-of-man-from- 14368906.php.

78 **because the service merely connected users to people:** Dryoff v. Ultimate Software Grp., 934 F.3d 1093, 1099 (9th Cir. 2019).

79 **pledge to eliminate AI-generated material of this kind:** Emma Woollacott, *Tech Firms Pledge to Eliminate AI-Generated CSAM*, Forbes (Apr. 24, 2024), https://www.forbes.com/sites/emmawoollacott/2024/04/24/tech-firms-pledge-to-eliminate-ai-generated-csam/.

79 **collaborate on techniques for flagging and taking down violent:** Paresh Dave, *Inside Two Years of Turmoil at Big Tech's Anti-Terrorism Group*, Wired (Sept. 30, 2024), https://www.wired.com/story/gifct-x-meta-youtube-microsoft-anti-terrorism-big-tech-turmoil/.

79 **have introduced friction into the ways in which their consumers share and engage:** *See, e.g.*, *Our Commitments*, YouTube, https://www.youtube.com/howyoutubeworks/our-commitments/curbing-extremist-content. *See generally* Ellen P. Goodman, Karen Kornbluh & Eli Weiner, *The Stakes of User Interface Design for Democracy*, German Marhsall Fund US (June 30, 2021), https://www.gmfus.org/publications/stakes-user-interface-design-democracy; Brett Frischmann & Susan Benesch, *Friction-In-Design Regulation as 21st Century Time, Place, and Manner Restriction*, 25 Yale J.L. & Tech. 376 (2023).

79 **free speech on platforms "is not the same as free reach":** Renee DiResta, *Free Speech Is Not the Same as Free Reach*, Wired (Aug. 30, 2018), https://www.wired.com/story/free-speech-is-not-the-same-as-free-reach/.

79 **such prompts dampen users' impulse to share harmful content:** See Garrett Morrow, Briony Swire-Thompson, Jessica M. Polny, Matthew Kopec & John P. Wihbey, *The Emerging Science of Content Labeling: Contextualizing Social Media Content Moderation* (Dec. 3, 2020) (unpublished manuscript), https://ssrn.com/abstract=3742120; Tom Dobber, Sanne Kruikemeier, Ellen P. Goodman, Natali Helberger & Sophie Minihold, *Effectiveness of Online Political Ad Disclosure Labels: Empirical Findings*, U. Amsterdam Inst. Info. L. (Mar. 8, 2021), https://www.uva-icds.net/wp-content/uploads/2021/03/Summary-transparency-discloures-experiment_update.pdf; Ellen Goodman & Karen Kornbluh, *The Stakes of User Interface Design for Democracy*, German Marshall Fund US (June 2021), https://www.gmfus.org/publications/stakes-user-interface-design-democracy; Ellen Goodman, *Digital Information Fidelity and Friction*, Knight First Amend. Inst. Colum. Univ. (Feb. 26, 2020), https://knightcolumbia.org/content/digital-fidelity-and-friction.

80 **moderation decisions on hate speech, harassment, nudity:** https://www.oversightboard.com/news/ban-on-showing-indigenous-nudity-disproportionately-limits-expression/; https://www.oversightboard.com/news/combat-misleading-deepfake-endorsements-by-changing-enforcement-approach/.

81 **would no longer amplify or demote posts about politics or political issues:** https://transparency.meta.com/features/approach-to-political-content.

81 **used to track misinformation on Facebook and Instagram:** Sarah Grevy Gotfredsen and Kaitlyn Dowling, *Meta Is Getting Rid of*

CrowdTangle—and Its Replacement Isn't as Transparent or Accessible, Columbia Journalism Review (July 9, 2024), https://www.cjr.org/tow_center/meta-is- getting-rid-of-crowdtangle.php.

81 **a lawsuit alleging that the boycott was a collusive conspiracy:** *World Federation of Advertisers Discontinues Small Brand Safety Initiative After Elon Musk's X Sues,* Associated Press (Aug. 9, 2024), https://apnews.com/article/x-twitter-advertiser-lawsuit-garm-discontinues-e64b99a501d9f69abbeebc1c479688d7.

CHAPTER FIVE

83 **"widen and accelerate the inextricable cycle of engagement › extraction › prediction › revenue":** Shoshana Zuboff, *Surveillance Capitalism or Democracy? The Death Match of Institutional Orders and the Politics of Knowledge in Our Information Civilization,* Organization Theory, 3(3) (2022). https://doi.org/10.1177/26317877221129290. See also Tim Wu, The Attention Merchants (Atlantic, 2016); Stefana Broadbent, Florian Forestier, Medhi Khamassi & Celia Zolynski, Pour une nouvelle culture de l'attention: Que faire de ces reseaux sociaux que nous épuisent? (Odile Jacob, 2024).

86 **white and suburban:** *See generally* Tanner Howard, *How GeoCities Suburbanized the Internet,* Bloomberg (Jan. 22, 2019), https://www.bloomberg.com/news/articles/2019-01-22/remembering-geocities-the-suburbia-of-the-early-web.

86 **its $4.95 per month subscription rate:** *See GeoCities in 1996,* Web Design Museum (last visited Mar. 18, 2024), https://www.webdesignmuseum.org/gallery/geocities-1996 (screenshots of GeoCities pages featuring advertisements for GeoPlus).

86 **keenly interested in a platform that could promote products and services:** *See* Benj Edwards, *Remembering GeoCities, the 1990s Precursor to Social Media,* How-To Geek (Oct. 3, 2020), https://www.howtogeek.com/692445/remembering-geocities-the-1990s-precursor-to-social-media/.

86 **users could buy GeoCities-branded merchandise:** *See GeoCities in 1998* (screenshot of the GeoCities homepage linking to the Marketplace and advertising a CyberShop).

86 **acquired GeoCities a few months later for $3.6 billion:** *See Yahoo Buys GeoCities,* CNN Money (January 28, 1999, 9:51 AM), https://money.cnn.com/1999/01/28/technology/yahoo_a/.

87 **Yahoo shut down the service in the US for good:** GeoCities survived under Yahoo in Japan until

2019. *See* Saqib Shah, *Yahoo Japan Is Shutting Down the Last Remnants of GeoCities*, Yahoo Money (Oct. 2, 2018), https://money.yahoo.com/2018-10-02-yahoo-japan-geocities-shutting-down.html.

87 **"one more about connecting people to people than people to websites":** Max Chafkin, *How to Kill a Great Idea! Jonathan Abrams Created the First Online Social Network and Enlisted Silicon Valley's Best and Brightest to Run It. Yet Friendster Flamed out Spectacularly. What Went Wrong?* Inc. (June 2007).

88 **the two companies each grew to about 115 million users:** Jonathan H. Kantor, *What Happened to MySpace and Who Owns It Now?* Slashgear, May 19, 2024, https://www.slashgear.com/1582318/what-happened-to-myspace-who-owns-it-now/.

89 **an online student directory through which users could share personal updates, photos, and video:** An earlier version, Facemash, invited users to pick the most attractive people on the site. The student founders dropped this project when Harvard threatened to expel them for it.

89 **it made the service feel like it had "a soul":** Farhad Manjoo, *Facebook News Feed Changed Everything*, Slate (Sept. 12, 2013), https://slate.com/technology/2013/09/facebook-news-feed-turns-7-why-its-the-most-influential-feature-on-the-internet.html.

90 **experimenting with and deploying personalized recommender systems:** https://www.amazon.science/tag/collaborative-filtering.

91 **to attend to its executives' shifting priorities and strategic concerns:** *See, e.g.*, Vladislav Vorotilov & Ilnur Shugaepov, *Scaling the Instagram Explore Recommendations System*, Facebook (Aug. 9, 2023), https://engineering.fb.com/2023/08/09/ml-applications/scaling-instagram-explore-recommendations-system/; Akos Lada, Meihong Wang & Tak Yan, *How Machine Learning Powers Facebook's News Feed Ranking Algorithm*, Facebook (Jan. 26, 2021), https://engineering.fb.com/2021/01/26/ml-applications/news-feed-ranking/.

91 **"censorship industrial complex":** Senate Hearing (2025), https://www.judiciary.senate.gov/committee-activity/hearings/the-censorship-industrial-complex; House Report (2024), https://judiciary.house.gov/sites/evo-subsites/republicans-judiciary.house.gov/files/evo-media-document/Biden-WH-Censorship-Report-final.pdf.

92 **across various interests like fashion, books, and home decor:** Simone Baribeau, *The Pinterest Pivot*, FastCompany (Oct. 23 2012),

 https://www.fastcompany.com/3001984/pinterest-pivot.

92 **an "Explore Feed" that highlights trending ideas:** Pinterest also offers a reverse image search tool called Lens, which allows users to find pins based on an uploaded photo. As of February 2018, Lens facilitated over six hundred million monthly searches. https://www.engadget.com/2018-02-08-pinterest-lens-visual-search-text-keywords.html.

92 **enabled companies to promote their products and services within feeds:** https://blog.hubspot.com/blog/tabid/6307/bid/33839/pinterest-finally-rolls-out-business-accounts-how-to-set-yours-up-today.aspx; https://www.engadget.com/2018-10-16-pinterest-product-pins.html; https://businessmodelanalyst.com/pinterest-business-model/. In 2022, Pinterest partnered with Warner Music Group and others to launch Idea Pins, which allow users to create multimedia pins featuring video, images, music, and text. https://techcrunch.com/2022/10/19/pinterest-partners-with-record-labels-to-bring-popular-music-to-its-tiktok-rival-idea-pins/.

92 **it targets ads and partners with influencers and merchants:** *The Pinterest Business Model*, Business Model Analyst (April 14, 2023), https://businessmodelanalyst.com/pinterest-business-model/.

92 **Pinterest went public in April 2019:** Maureen Farrell, *Pinterest Files Confidentially for IPO*, Wall St. J. (updated Feb. 21, 2019), https://www.wsj.com/articles/pinterest-files-confidentially-for-ipo-11550782552.

92 **most of whom are women:** https://business.pinterest.com/audience/.

92 **through which users could play games with friends:** https://techcrunch.com/2016/11/29/messenger-instant-games/; https://www.lifewire.com/what-is-facebook-3486391.

93 **the most popular networking platform for professionals:** *LinkedIn*, Britannica Money, https://www.britannica.com/money/LinkedIn.

93 **young professionals between the ages of twenty-five and thirty-four:** Jacqueline Zote, *28 LinkedIn Statistics That Marketers Must Know in 2025*, Sprout Social (Apr. 4, 2025), https://sproutsocial.com/insights/linkedin-statistics/.

94 **distributes ad-supported short-form content from major publishers:** Amit Chowdhry, *Snapchat's New "Discover" Feature Has Content from ESPN, CNN, Food Network and Others*, Forbes (Jan. 28, 2015), https://www.forbes.com

/sites/amitchowdhry/2015/01/28/snapchat-discover/.

94 **location-based "geofilters" with corporate sponsors:** https://web.archive.org/web/20181118162852/https://forbusiness.snapchat.com/; https://www.ibtimes.com/snapchats-super-bowl-sponsored-lens-gatorade-had-more-100-million-views-2298865; https://www.adweek.com/performance-marketing/twentieth-century-fox-buys-first-snapchat-lens-takeover-ad-171639/.

94 **the platform reaches 453 million daily active users:** https://investor.snap.com/news/news-details/2025/Snap-Inc.-Announces-Fourth-Quarter-and-Full-Year-2024-Financial-Results/default.aspx.

95 **to combine visual content-sharing with location-based commercial surveillance techniques:** https://www.theatlantic.com/technology/archive/2012/10/how-instagram-beat-hipstamatic-its-own-game/322743/; https://www.investopedia.com/articles/investing/102615/story-instagram-rise-1-photo sharing-app.asp#citation-14.

95 **when Facebook acquired it for $1 billion:** https://about.instagram.com/about-us/instagram-product-evolution; https://techcrunch.com/2012/03/11/instagram-reaches-27-million-registered-users-shows-off-upcoming-android-app/.

95 **This rawness fostered a sense of license and authenticity:** Josh Halliday, *Twitter's Tony Wang: "We Are the Free Speech Wing of the Free Speech Party,"* The Guardian (Mar. 22, 2012), https://www.theguardian.com/media/2012/mar/22/twitter-tony-wang-free-speech.

96 **Zuckerberg said, "We run ads":** Dylan Byers, *Senate Fails Its Zuckerberg Test*, CNN (Apr. 11, 2018), https://money.cnn.com/2018/04/10/technology/senate-mark-zuckerberg-testimony/index.html.

96 **than they would if they negotiated for ad placements:** *See generally* European Commission, Directorate-General for Communications Networks, Content and Technology, C. Armitage, N. Botton, L. Dejeu-Castang et al., *Towards a More Transparent, Balanced and Sustainable Digital Advertising Ecosystem: Study on the Impact of Recent Developments in Digital Advertising on Privacy, Publishers and Advertisers—Final Report* 112 & 115 (2023), https://op.europa.eu/en/publication-detail/-/publication/8b950a43-a141-11ed-b508-01aa75ed71a1/language-en.

96 **"also more likely to be associated with lower quality vendors and higher product**

 prices": Eduardo Schnadower Mustri, Idris Adjerid & Allesandro Acquisti, *Behavioral Advertising and Consumer Welfare* (Apr. 2023; revised Apr. 2024), https://papers.ssrn.com/sol3/papers.cfm?abstract_id=4398428.

97 **"liked the product's political message more than the razor itself":** Julia Angwin, *If It's Advertised to You Online, You Probably Shouldn't Buy It. Here's Why.*, N.Y. Times (Apr. 6, 2023), https://www.nytimes.com/2023/04/06/opinion/online-advertising-privacy-data-surveillance-consumer-quality.html.

97 **widespread discrimination based on gender, age, and race:** Julia Angwin & Terry Parris Jr., *Facebook Lets Advertisers Exclude Users by Race*, ProPublica (Oct. 28, 2016), https://www.propublica.org/article/facebook-lets-advertisers-exclude-users-by-race. *See generally* Sylvain, *Discriminatory Designs*; Sylvain, *Intermediary Duties*; Gillian B. White, *When Algorithms Don't Account for Civil Rights*, The Atlantic (Mar. 7, 2017), https://www.theatlantic.com/business/archive/2017/03/facebook-ad-discrimination/518718.

97 **only after being alerted to it by journalists:** Jeremy B. Merrill, *Google Has Been Allowing Advertisers to Exclude Nonbinary People from Seeing Job Ads*, Markup (Feb. 11, 2021, 8:00 AM), https://themarkup.org/google-the-giant/2021/02/11/google-has-been-allowing-advertisers-to-exclude-nonbinary-people-from-seeing-job-ads.

97 **ad tech market's opacity has also enabled fraud:** Nicola Agius, *$84 Billion of Ad Spend Lost Due to Ad Fraud in 2023*, Search Engine Land (Sept. 28, 2023), https://searchengineland.com/ad-spend-lost-ad-fraud-2023-432610; Paula Chiocchi, *Ad Fraud: The Biggest Threat to the Advertising Industry*, Forbes (Nov. 7, 2023), https://www.forbes.com/councils/forbesagencycouncil/2023/11/07/ad-fraud-the-biggest-threat-to-the-advertising-industry/. *See also* Roberto Cavazos, The Economic Cost of Bad Actors on the Internet: Ad Fraud (2020), https://info.cheq.ai/hubfs/Research/Economic-Cost-BAD-ACTORS-ON-THE-INTERNET-Ad-Fraud-2020.pdf.

97 **they sometimes provide faulty or incomplete data:** *See, e.g.*, Mark Scott, *Thousands of Posts Around January 6 Riots Go Missing from Facebook Transparency Tool*, Politico (Aug. 31, 2021), https://www.politico.eu/article/facebook-crowdtangle-data-january-6-capitol-hill-riots- transparency/; Davey Alba, *Facebook Sent Flawed Data to Misinformation Researchers*, N.Y. Times (Sept. 10, 2021), https://www.nytimes.com/live/2020/2020-election-misinformation

-distortions#facebook-sent-flawed-data-to-misinformation-researchers.

97 **inflate or otherwise mischaracterize the effectiveness:** *See, e.g.*, Gilad Edelman, *Ad Tech Could Be the Next Internet Bubble*, WIRED (Oct. 5, 2020), https://www.wired.com/story/ad-tech-could-be-the-next-internet-bubble/; *The Risks and Rewards of Targeted Ads*, YALE SCH. OF MGMT. BLOG (May 8, 2020), https://som.yale.edu/blog/the-risks-and-rewards-of-targeted-ads; Erika Wheless, *How a Screensaver Cheated Connected TV Advertisers Out of $6 Million*, AD AGE (Aug. 12, 2021), https://adage.com/article/digital-marketing-ad-tech-news/how-screensaver-cheated-connected-tv-advertisers- out-6-million/2357606; Parmy Olson, *New Tactics Punch Holes in Big Tech's Ad-Fraud Defenses*, WALL. ST. J. (Jan. 7, 2020), https://www.wsj.com/articles/new-tactics-punch-holes-in-big-techs-ad-fraud-defenses-11578418893?mod=article_relatedinline.

98 **because of the sheer scope and size of their potential consumer reach:** *See* European Commission, Directorate-General for Communications Networks, Content and Technology, C. Armitage, N. Botton, L. Dejeu-Castang. et al., *Towards a More Transparent, Balanced and Sustainable Digital Advertising Ecosystem: Study on the Impact of Recent Developments in Digital Advertising on Privacy, Publishers and Advertisers—Final Report* 121–22 (2023), https://op.europa.eu/en/publication-detail/-/publication/8b950a43-a141-11ed-b508-01aa75ed71a1/language-en.

98 **overstated the performance of video ads:** Steven Perlberg, *Facebook Apologizes for Video Metric Miscalculation*, WALL. ST. J. (Sept. 23, 2016), https://www.wsj.com/articles/facebook-apologizes-for-video-metric-miscalculation-1474641054.

98 **revealed that it overstated video views and ad impressions:** Sahil Patel, *LinkedIn Finds Measurement Errors That Inflated Video and Ad Metrics*, WALL. ST. J. (Nov. 12, 2020), https://www.wsj.com/articles/linkedin-finds-measurement-errors-that-inflated-video-and-ad-metrics-11605228577.

98 **case settled for $100 million in 2025:** Jonathan Stempel, *Google to Pay $100 Million to Settle Advertisers' Class Action*, REUTERS (Mar. 28, 2025), https://www.reuters.com/legal/google-pay-100-million-settle-advertisers-class-action-2025-03-28/.

99 **overcharging based on those misleading metrics:** Winston Cho, *Google Sued by Advertisers for Allegedly Inflating Video Ad Metrics*, HOLLYWOOD REPORTER (July 26, 2023),

https://www.hollywoodreporter.com/business/business-news/google-lawsuit-inflating-video-ad-metrics-1235546090/.

99 **retaliated against firms that sought to challenge:** Karina Montoya, *Five Takeaways from the DOJ's Case in the Google Ad Tech Trial*, Tech Policy Press (Sept. 23, 2024), https://www.techpolicy.press/five-takeaways-from-the-dojs-case-in-the-google-ad-tech-trial/.

99 **against data brokers who unfairly traffic in consumers' sensitive data:** *See* Federal Trade Commission, *FTC Sues Kochava for Selling Data That Tracks People at Reproductive Health Clinics, Places of Worship, and Other Sensitive Locations* (Aug. 29, 2022), https://www.ftc.gov/news-events/news/press-releases/2022/08/ftc-sues-kochava-selling-data-tracks-people-reproductive-health-clinics-places-worship-other; Federal Trade Commission, Press Release, *FTC Order Prohibits Data Broker X-Mode Social and Outlogic from Selling Sensitive Location Data* (Jan. 9, 2024), https://www.ftc.gov/news-events/news/press-releases/2024/01/ftc-order-prohibits-data-broker-x-mode-social-outlogic-selling-sensitive-location-data; Federal Trade Commission, Press Release, *FTC Order Will Ban InMarket from Selling Precise Consumer Location Data* (Jan. 18, 2024), https://www.ftc.gov/news-events/news/press-releases/2024/01/ftc-order-will-ban-inmarket-selling- precise-consumer-location-data.

99 **from "real-time bidding" exchanges:** Federal Trade Commission, Press Release, *FTC Takes Action Against Mobilewalla for Collecting and Selling Sensitive Location Data* (Dec. 3, 2024), https://www.ftc.gov/news-events/news/press-releases/2024/12/ftc-takes-action-against-mobilewalla-collecting-selling-sensitive-location-data.

100 **because they bring so much near-term gain:** Tom Cunningham, Sana Pandey, Leif Sigerson, Jonathan Stray, Jeff Allen, Bonnie Barrilleaux, Ravi Iyer, Smitha Milli, Mohit Kothari & Behnam Rezaei, What We Know About Using Non-Engagement Signals in Content Ranking (Feb. 9, 2024), https://arxiv.org/abs/2402.06831.

100 **during the heyday of traditional broadcast television or print publishing:** *See* European Commission, Directorate-General for Communications Networks, Content and Technology, C. Armitage, N. Botton, L. Dejeu-Castang et al., *Towards a More Transparent, Balanced and Sustainable Digital Advertising Ecosystem: Study on the Impact of Recent Developments in Digital Advertising on Privacy, Publishers and Advertisers—Final Report* 114

(2023), https://op.europa.eu/en/publication-detail/-/publication/8b950a43-a141-11ed-b508-01aa75ed71a1/language-en.

100 **Jell-O mold recipes, or right-wing militia gear:** *See, e.g.*, Meghan Graham, *Why That Ad for Butt-Flap Pajamas Is Following You All Over the Internet*, CNBC (Dec. 21, 2020), https://www.cnbc.com/2020/12/21/why-that-ad-for-butt-flap-pajamas-is-following-you-all-over-the-internet.html; Steven Melendez, *Feeling Brave? Try These Disturbing Jell-O Mold Recipes Created by a Bot*, Fast Company (Feb. 7, 2020), https://www.fastcompany.com/90462023/feeling-brave-try-these-disturbing-jell-o-mold-recipes-created-by-a-bot; Jeremy B. Berrill, *Tech Firms Profited from Far-Right Militia Content Despite Ban on "Three Percenters,"* Markup (Jan. 21, 2021), https://themarkup.org/news/2021/01/21/tech-firms-profited-from-far-right-militia-content-despite-ban-on-three-percenters.

100 **content quality measures like "authoritativeness":** Justin Hendrix, *Ranking Content on Signals Other Than User Engagement*, Tech Policy Press (Feb. 18, 2024), https://www.techpolicy.press/ranking-content-on-signals-other-than-user-engagement/.

101 **contextual models serve ads based on the content of the recipe:** Some researchers have found that contextual advertising may be as cost-effective as targeting, if not more so. *See* Veronica Marotta, Vibhanshu Abhishek & Alessandro Acquisti, *Online Tracking and Publishers' Revenues: An Empirical Analysis* (2019), https://weis2019.econinfosec.org/wp-content/uploads/sites/6/2019/05/WEIS_2019_paper_38.pdf); Veronica Marotta, Kaifu Zhang & Allesandro Acqusti, *The Welfare Impact of Targeted Advertising Technologies*, Information Systems Research (2021) (published version of Who Benefits from Targeted Advertising, https://www.heinz.cmu.edu/~acquisti/papers/Acquisti_Welfare_Impact_of_Targeted_Advertising_WP.pdf).

102 **"something of a conscientious objector":** Mike Allen, *Sean Parker Unloads on Facebook: "God Only Knows What It's Doing to Our Children's Brains,"* Axios (Nov. 9, 2017).

103 **content that diminished the self-esteem of young women:** See Frances Haugen, The Power of One: How I Found the Strength to Tell the Truth and Why I Blew the Whistle on Facebook (Little, Brown & Co., 2023).

103 **company's unmitigated ambition for consumer attention:** See Sarah Wynn-Williams, Careless People: A Cautionary Tale of Power, Greed, and Lost Idealism (Macmillan, 2025).

103 **"unpredictable rewards on schedules structured to exploit human psychology":** Matthew B. Lawrence, *Supreme Court Endorses Neutrality Triangulation Approach to Constitutionality of Platform Regulation*, BALKINIZATION (August 22, 2024), https://balkin.blogspot.com/2024/08/gb05.html.

103 **when they feel that they have earned an unexpected reward:** See generally Natasha Dow Schull, ADDICTION BY DESIGN: MACHINE GAMBLING IN LAS VEGAS (Princeton University Press, 2012).

104 **"may be particularly vulnerable to technological addiction":** American Psychiatry Association, *Media Advisory: As a Third of Americans Spend Four or More Hours a Day on Social Media, APA Offers New Polling*, RESOURCES ON TECHNOLOGY USE (Apr. 10, 2024), https://www.psychiatry.org/news-room/news-releases/media-advisory-resources-on-technology-use.

104 **same way that tobacco companies once marketed cigarettes to teens:** Monique Merrill, *California Pushes Ninth Circuit to Lift Block on Its Law Protecting Children from Social Media Addiction*, COURTHOUSE NEWS SERVICE (Apr. 2, 2025), https://www.courthousenews.com/california-pushes-ninth- circuit-to-lift-block-on-its-law-protecting-children-from-social-media-addiction/.

104 **fostering unprecedented levels of anxiety and depression:** Jonathan Haidt, THE ANXIOUS GENERATION (Penguin Random House, 2024).

104 **a 2023 surgeon general's advisory on Social Media and Youth Mental Health:** US Surgeon General, *Social Media and Youth Mental Health* (2023), https://www.hhs.gov/sites/default/files/sg-youth-mental-health-social-media-advisory.pdf.

104 **"to be defined by user-created content":** Chris Cox, https://www.thecgo.org/research/section-230-a-retrospective/#congressional-intent-in-practice-how-section-230-works; Ron Wyden, https://www.wyden.senate.gov/news/press-releases/wyden-remarks-at-section-230-briefing-hosted-by-eff.

CHAPTER SIX

108 **They are winning some of these cases:** *See, e.g.*, Alex Picket, *Judge Halts Florida's Social Media Ban for Minors*, COURTHOUSE NEWS SERVICE (June 3, 2025), https://www.courthousenews.com/judge-halts-floridas-social-media-ban-for-minors/.

108 **But they are losing others:** State of Vermont v. Meta Platforms, Inc., 2024 WL 3741424 (Vt. Super.

Ct. July 29, 2024); Suffolk v. Meta Platforms, Inc., 2024 WL 464835 (Mass. Super. Ct. Suffolk Cty. Oct. 17, 2024).

109 **"Goldilocks zone":** Alfred Ng, *The Shifting Sands of Data Privacy Law*, Politico (June 25, 2025), https://www.politico.com/newsletters/digital-future-daily/2025/06/25/the-shifting-sands-of-national-data-privacy-law-00423795.

109 **that it uses to fine brick-and-mortar fraudsters:** *See, e.g.*, Federal Trade Commission, Press Release, *Twitter to Pay $150 Million Penalty for Allegedly Breaking Its Privacy Promises—Again* (May 25, 2022), https://www.ftc.gov/news-events/news/press-releases/2022/05/ftc-charges-twitter-deceptively-using-account-security-data-sell-targeted-ads; Federal Trade Commission, *Evoke Wellness to Pay $1.9 Million to Settle FTC Claims That They Misled Consumers Seeking Substance Use Disorder Treatment* (June 10, 2025), https://www.ftc.gov/news-events/news/press-releases/2025/06/evoke-wellness-pay-19-million-settle-ftc-claims-they-misled-consumers-seeking-substance-use-disorder.

109 **hiding the early termination fee for its most popular subscription plan:** Federal Trade Commission, Press Release, *FTC Takes Action Against Adobe and Executives for Hiding Fees, Preventing Consumers from Easily Cancelling Software Subscriptions* (June 17, 2024), https://www.ftc.gov/news-events/news/press-releases/2024/06/ftc-takes-action-against-adobe-executives-hiding-fees-preventing-consumers-easily-cancelling.

109 **unintended in-game purchases without their parents' permission:** Federal Trade Commission, Press Release, *Fortnite Video Game Maker Epic Games to Pay More Than Half a Billion Dollars over FTC Allegations of Privacy Violations and Unwanted Charges* (Dec. 19, 2022), https://www.ftc.gov/news-events/news/press-releases/2022/12/fortnite-video-game-maker-epic-games-pay-more-half-billion-dollars-over-ftc-allegations.

110 **a toxic mix for all adolescents:** Federal Trade Commission, Press Release, *FTC Order Will Ban NGL Labs and Its Founders from Offering Anonymous Messaging Apps to Kids Under 18 and Halt Deceptive Claims Around AI Content Moderation* (July 9, 2024), https://www.ftc.gov/news-events/news/press-releases/2024/07/ftc-order-will-ban-ngl-labs-its-founders- offering-anonymous-messaging-apps-kids-under-18-halt.

110 **"real-time bidding" exchanges:** Federal Trade

Commission, Press Release, *FTC Takes Action Against Mobilewalla for Collecting and Selling Sensitive Location Data* (Dec. 3, 2024), https://www.ftc.gov/news-events/news/press-releases/2024/12/ftc-takes-action-against-mobilewalla-collecting-selling-sensitive-location-data.

111 **who unfairly trafficked in consumers' sensitive data:** *See* Federal Trade Commission, *FTC Sues Kochava for Selling Data That Tracks People at Reproductive Health Clinics, Places of Worship, and Other Sensitive Locations* (Aug. 29, 2022), https://www.ftc.gov/news-events/news/press-releases/2022/08/ftc-sues-kochava-selling-data-tracks-people-reproductive-health-clinics-places-worship-other; Federal Trade Commission, Press Release, *FTC Order Prohibits Data Broker X-Mode Social and Outlogic from Selling Sensitive Location Data* (Jan. 9, 2024), https://www.ftc.gov/news-events/news/press-releases/2024/01/ftc-order-prohibits-data-broker-x-mode-social-outlogic-selling-sensitive-location-data; Federal Trade Commission, Press Release, *FTC Order Will Ban InMarket from Selling Precise Consumer Location Data* (Jan. 18, 2024), https://www.ftc.gov/news-events/news/press-releases/2024/01/ftc-order-will-ban-inmarket-selling-precise-consumer-location-data.

114 **were *actually* liable for terrorist attacks around the world:** Twitter, Inc. v. Taamneh, 598 US 471 (2023); Gonzalez v. Google, LLC, 598 US 617 (2023).

115 **the opportunity to narrow the scope of the protection under Section 230:** Malwarebytes; Gonzalez/Taamneh; Snap (2024).

115 **"Social-media platforms have increasingly used § 230 as a get-out-of-jail free card":** Doe through Roe v. Snap Inc., 144 S.Ct. 2493, 2494 (July 2, 2024) (Thomas, dissenting from denial of certiorari).

116 **Roommates, a website that matches potential co-tenants:** Fair Housing Council of San Fernando Valley v. Roommates.com, 521 F.3d 1157 (9th Cir. 2008).

117 **what Congress sought to encourage with Section 230:** The Ninth Circuit remanded the case back to the trial court to determine whether, through the drop-down menu, Roommates violated federal and state housing laws now that the liability shield was not applicable. *Id.* at 1175. In the end, the trial court determined that it did not. Fair Hous. Council of San Fernando Valley v. Roommates.com, 666 F.3d 1216, 1219 (9th Cir. 2012).

117 **second-most-cited Section 230 case after *Zeran*:** As of the end of 2024, 415 cases have cited *Roommates* and 430 have cited

Zeran. A couple other cases that cite Section 230 have more citations, but for reasons that are unrelated to Section 230, as such.

117 **whether a defendant is an "information content provider":** See Valerie C. Brannon & Eric N. Holmes, Cong. Rsch. Serv., R46751, Section 230: An Overview 18 (2024), https://crsreports.congress.gov/product/pdf/R/R46751.

117 **new way by which plaintiffs could overcome the liability shield:** I hopefully speculated at the possibility in my first article on the statute. Olivier Sylvain, *Intermediary Design Duties*, 50 Conn. L. Rev. 203 (2018).

118 **and then posted that material on the website:** Jones v. Dirty World Entertainment Recordings, 755 F.3d 398 (6th Cir. 2014).

118 **holding that a third party, Herrick's ex-boyfriend:** Herrick v. Grindr, 765 Fed.Appx. 586, 590 (2d Cir. 2019).

119 **materially supported Hamas operatives:** Force v. Facebook, 934 F.3d 53, 59 (2d Cir. 2019).

119 **"more visible, available, and usable":** Force v. Facebook, Inc., 934 F.3d 53, 70 (2d Cir. 2019).

119 **"friend- and content-suggestion algorithms":** *Id.* at 82 (Katzmann, C.J., concurring in part and dissenting in part). He concurred with rest of the majority's opinion.

120 **ways in which platforms shape consumers' online experiences:** It was also notable because of who he was; judges, policymakers, and academics generally admired Judge Katzmann's writing on statutory interpretation and the relationship between the courts and Congress. See Robert A. Katzmann, Courts and Congress (Brookings Institute Press, 1997); Robert A. Katzmann, *Madison Lecture: Statutes*, 87 NYU L. Rev. 637 (2012); Robert A. Katzmann, Judging Statutes (Oxford, 2014).

120 **"It strains the English language":** Malwarebytes, Inc. v. Enigma Software Grp. USA, LLC, 141 S. Ct. 13, 17 (2020) (Thomas, J., statement respecting denial of certiorari). The standalone statement was itself a kind of oddity, since it was neither a dissent nor a concurrence. Justice Thomas returned to this view in yet another opinion, this time a true dissent from the majority's decision to deny review in a case involving Snap. See Doe through Roe v. Snap, Inc., 44 S.Ct. 2493 (July 2, 2024) (Thomas, J., dissenting from denial of certiorari).

121 **not Snap as a publisher:** *Id.* at 1093–94.

121 **matching algorithms are shielded under Section 230:** VV v. Meta Platforms, 2023 WL 3613232 (Conn. 2024).

122 **Omegle was defective under product liability and negligent design theories:** A.M. v. Omegle, 614 F.Supp.3d 814, 817 (D. Or. 2022). She also claimed that the service unlawfully facilitated sex and human trafficking.

122 **against Facebook, Instagram, YouTube, TikTok, and Snapchat:** In re Social Media Adolescent Addiction/Personal Injury Products Liability Litigation, 753 F.Supp.3d 849 (N.D. Cal. 2024).

123 **such claims are squarely about the design of the platform:** In re Social Media Adolescent Addiction/Personal Injury Products Liability Litigation, 702 F.Supp.3d 809, 833 (N.D. Cal. 2023).

123 **owns a patent for a system that determines the best times to send notifications:** US Patent No. 8,751,636 B2 (filed Dec. 22, 2010) (issued June 10, 2014).

124 **"their own expressive activity or content (i.e., first-party speech)":** Anderson v. TikTok, Inc., 116 F.4th 180, 183 (3d. Cir. 2024).

CHAPTER SEVEN

125 **an exception to the liability shield for sex trafficking:** Public Law 115-164 (2018) (Allow States and Victims to Fight Online Sex Trafficking Act & Stop Enabling Sex Traffickers Act [FOSTA-SESTA]).

125 **criminalizes the nonconsensual publication of intimate visual depictions:** Public Law 119-146 (2025) (Tools to Address Known Exploitation by Immobilizing Technological Deepfakes on Websites and Networks Act). This effort builds on the impressive state-by-state campaign of the Cyber Civil Rights Initiative and allied activists to criminalize nonconsensual pornography in all fifty states, the District of Columbia, and a couple US territories. https://cybercivilrights.org/nonconsensual-distribution-of-intimate-images/.

127 **land mines in the doctrine for content-based regulation:** Opponents of the new carve-out for sex trafficking argued that the change would chill voluntary moderation by websites and, relatedly, harm sex workers for whom it would be harder to find safe online resources. Opponents of the Take It Down Act alleged that the law would incentivize people to claim in bad faith that visual depictions, including satire involving political officials, must be blocked.

128 **was not whether the new administration would attend to the plutocrats:** Clara Ence Morse,

Cat Zakrzewski, Aaron Schaffer, Luis Melgar & Nick Mourtoupalas, *Meet the Top Donors to Trump's $239 Million Inauguration Fund,* Wash. Post (updated May 9, 2025), https://www.washingtonpost.com/politics/interactive/2025/trump-inauguration-donors-list/.

128 **on states and municipalities that implement their own AI regulations:** Theodoric Meyer & Maegan Vazquez, *The House Passed Trump's Major Budget Bill. What Happens Next?* Wash. Post (May 22, 2025), https://www.washingtonpost.com/politics/2025/05/22/trump-budget-bill-congress/.

128 **would have withheld funding to governments that do so:** The Senate Commerce Committee revised the bill to tie it more directly to budgeting, which the chamber's "Byrd rule" requires. The revision provides that funding for certain programs would be cut off to states or local governments that implement "any law or regulation limiting, restricting, or otherwise regulating artificial intelligence models, artificial intelligence systems, or automated decision systems entered into interstate commerce."

129 **have been eager to lift regulatory burdens on innovation:** Americans for Responsible Innovation, AI Law Preemption Report: Federal Preemption of AI Regulation: What State Legislation Is at Risk? at 2 (June 6, 2025), https://ari.us/wp-content/uploads/2025/06/ARI-Report-Fed-Preemption-6-6-25.pdf.

129 **have been eager to lift regulatory burdens on innovation:** Alex Rogers & Stephen Morris, *Big Tech Pushes for 10-Year Ban on US States Regulating AI,* Financial Times (June 18, 2025), https://www.ft.com/content/52ae52f1-531e-462f-898f-e9f86b3b1869.

129 **state laws are "disastrous" for AI development:** US Senate Committee on Commerce, Science, and Transportation, Press Release: Sen. Cruz: Adopting Europe's Approach on Regulation Will Cause China to Win the AI Race (May 8, 2025), https://www.commerce.senate.gov/2025/5/sen-cruz-adopting-europe-s-approach-on-regulation-will-cause-china-to-win-the-ai-race. See also Miranda Nazzaro, *Lawmakers Push Tech Leaders on AI, Energy in Race with China,* The Hill (May 8, 2025), https://thehill.com/policy/technology/5291184-lawmakers-push-tech-leaders-on-ai-energy-in-race-with-china/; Kevin Frazier & Adam Thierer, *1,000 AI Bills: Time for Congress to Get Serious About Preemption,* LawFare (May 9, 2025), https://www.lawfaremedia.org/article/1-000-ai-bills--time-for-congress-to-get-serious-about-preemption.

129 **"the nation's efforts to stay at the cutting edge of AI innovation":** Robert Boothe, *Trump's Plan to Ban US States From AI Regulation Will "Hold Us Back," Says Microsoft Science Chief,* The Guardian (June 22, 2025), https://www.theguardian.com/technology/2025/jun/22/trump-ban-us-states-ai-regulation-microsoft-eric-horvitz.

129 **extolled the states as the "laboratories of experimentation":** New State Ice Co. v. Liebmann, 285 US 262 (1932).

130 **an important check against federal power:** See AT&T Corp. v. Iowa Utilities Board, 525 US 366 (1999); Nixon v. Missouri Municipal League, 541 US 125 (2004).

130 **are concerned about incursions on states' rights:** Miranda Nazzaro, *AI Moratorium Sparks GOP Battle Over States' Rights,* The Hill (July 18, 2025), https://thehill.com/policy/technology/5355684-ai-moratorium-sparks-gop-battle-over-states-rights/.

131 **The current FTC Chair has even warned companies against complying:** Mila Fiordalisi, *The FTC Warns Big Tech Companies Not to Apply the Digital Services Act,* Wired (Aug. 31, 2025), https://www.wired.com/story/big-tech-companies-in-the-us-have-been-told-not-to-apply-the-digital-services-act/.

131 **other means of curbing the excesses of the big social media platforms:** Some thoughtful staff on the Hill have entertained such reforms, too, but these have not yet found the light of day in public hearings, floor colloquies, or press conferences.

131 **seek to foster purpose-driven local discussions:** Cf. Deepti Doshi, *Local Digital Spaces Are Essential Civic Infrastructure,* Stanford Social Innovation Rev. (Spring 2025), https://archive.is/gOSlw#selection-1565.149-1565.269. See also Ethan Zuckerman, *The Case for Digital Social Infrastructure,* Knight First Amendment Institute at Columbia (Jan. 17, 2020), https://knightcolumbia.org/content/the-case-for-digital-public-infrastructure.

131 **and a premium service with a few more options:** *How a Social Network Is Bringing People Together in Increasingly Divisive Times,* PBS Newshour (May 7, 2025), https://www.pbs.org/newshour/show/how-a-social-network-is-bringing-people-together-in-increasingly-divisive-times.

132 **embrace online moderation features that prioritize user governance:** See Ethan Zuckerman & Chand Rajendra-Nicolucci, *From Community*

Governance to Customer Service and Back Again: Re-Examining Pre-Web Models of Online Governance to Address Platforms' Crisis of Legitimacy, Social Media + Society (Sept. 8, 2023), https://doi.org/10.1177/20563051231196864. *See also* https://newpublic.org/.

132 **X's "Community Notes":** https://communitynotes.x.com/guide/en/about/introduction.

132 **work in much the same way that browser plug-ins for personalized privacy protection do:** The right to "data portability" in Article 20 of the EU's General Data Protection Regulation arises from the same end user focus. Through this right, consumers can travel from online service with their data to another service.

132 **AI-powered software to browse, evaluate, filter, sort, and format:** Mark Purdy, *What Is Agentic AI, and How Will It Change Work?* Harvard Business Review (Dec. 12, 2024), https://hbr.org/2024/12/what-is-agentic-ai-and-how-will-it-change-work.

133 **"Unfollow Everything 2.0":** Steve Dent, *Researcher's "Unfollow Everything" Lawsuit Against Meta Gets Dismissed,* Engadget (Nov. 8, 2024), https://www.engadget.com/social-media/researchers-unfollow-everything-lawsuit-against-meta-gets-dismissed-133051131.html.

133 **has threatened to sue people and researchers who use similar tools:** *See* Laura Edelson & Damon McCoy, *We Research Misinformation on Facebook. It Just Disabled Our Accounts.,* N.Y. Times (Aug. 10, 2021), https://www.nytimes.com/2021/08/10/opinion/facebook-misinformation.html; Louis Barclay, *Facebook Banned Me for Life Because I Help People Use It Less,* Slate (Oct. 7, 2021), https://slate.com/technology/2021/10/facebook-unfollow-everything-cease-desist.html.

133 **immunizes the use of technical means for filtering offensive user-generated content:** 47 USC. § 230(c)(2)(B).

133 **dismissed the case because no one has actually used the tool:** Zuckerman v. Meta Platforms, 2024 WL 4876949 (N.D. Cal. Nov. 22, 2024).

133 **can only be as effective as tech-focused researchers like Zuckerman:** Isabel Adler, *Zuckerman v. Meta: A User-Friendly Section 230,* Knight First Amendment Institute at Columbia (Oct. 11, 2024), https://knightcolumbia.org/blog/zuckerman-v-meta-a-user-friendly-section-230-1.

133 **as it appears many aspire to do:** According to Statista, about 912 million people used Google Chrome's AdBlock plugin. Statista, *Number of Adblock Users Worldwide*

in Selected Quarters From 2012 to 2023, https://www.statista.com/statistics/435252/adblock-users-worldwide/ (accessed on Dec. 10, 2024).

134 **a composable user experience:** Luke Hogg & Renée DiResta, *Shaping the Future of Social Media with Middleware*, Foundation for American Innovation (Dec. 17, 2024), https://www.thefai.org/posts/shaping-the-future-of-social-media-with-middleware.

134 **"dramatically dilute the power of the platforms to amplify fringe views":** Fukuyama at p. 43.

134 **most notable of such protocols among a new crop of social networking platforms:** Meanwhile, billionaire entrepreneur Frank McCourt is spearheading the development of a whole new common protocol (the "Decentralized Social Networking Protocol") that prioritizes user control over their identity and personal information. See https://www.projectliberty.io/about/. Project Liberty, the organization that is managing McCourt's project, submitted a "People's Bid" to buy TikTok after Congress required the latter to cut all ties with the China-based holding company, ByteDance, in 2025. See https://www.thepeoplesbid.com/.

134 **"push the power and decision-making out to the ends of the network":** Mike Masnick, *Protocols, Not Platforms: A Technological Approach to Free Speech*, Knight First Amendment Institute at Columbia University (Aug. 21, 2019), https://knightcolumbia.org/content/protocols-not-platforms-a-technological-approach-to-free-speech.

134 **"erodes the high walls that monopolies build to protect themselves":** Bennett Cyphers & Cory Doctorow, *A Legislative Path to an Interoperable Internet*, Electronic Frontier Foundation (July 28, 2020), https://www.eff.org/deeplinks/2020/07/legislative-path-interoperable-internet. *See also* Richard Reisman, *Three Pillars of Human Discourse (and How Social Media Middleware Can Support All Three)*, Tech Policy Press (Oct. 24, 2024), https://www.techpolicy.press/three-pillars-of-human-discourse-and-how-social-media-middleware-can-support-all-three/.

134 **mandating legally enforceable interoperability standards:** To be fair, this is precisely what antitrust advocates have been urging in the context of digital markets, without necessarily having middleware in mind. Subcommittee on Antitrust, Commercial, and Administrative Law, Judiciary Committee, House of Representatives, Investigation of Competition in Digital Markets (adopted April 2021), https://www.govinfo.gov/content/pkg

/CPRT-117HPRT47832/pdf /CPRT-117HPRT47832.pdf; Herbert Hovenkamp, *Antitrust Interoperability Remedies*, 123 Colum. L. Rev. F. 1 (2023), https://columbialawreview.org/content/antitrust-interoperability-remedies/.

CONCLUSION

138 **Congress could repeal or sunset Section 230 altogether:** Cathy McMorris Rodgers & Frank Pallone Jr., *Sunset of Section 230 Would Force Big Tech's Hand*, Wall St. J. (May 12, 2024), https://www.wsj.com/articles/sunset-of-section-230-would-force-big-techs-hand-208f75f1.

138 **there is good reason to doubt that a total repeal would be cataclysmic:** Baidu; Prager.

139 **Congress would probably do better:** FCC Commissioner Brendan Carr has said that, under his leadership, the agency will promulgate an interpretation of the statute pursuant to its rulemaking authority. Avery Lotz, *What to Know About Brendan Carr, Trump's Pick for FCC Chair*, Axios (Nov. 18, 2024), https://www.axios.com/2024/11/18/brendan-carr-trump-fcc-chair. This is something the first Trump administration and the FCC Chair at the time attempted but never actually came anywhere close to accomplishing. *See* White House, Executive Order on Preventing Online Censorship (May 28, 2020), https://trumpwhitehouse.archives.gov/presidential-actions/executive-order-preventing-online-censorship/. *See also* Thomas M. Johnson, *The FCC's Authority to Interpret Section 230 of the Communications Act*, FCC Blog (Oct. 21, 2020) (statement of FCC General Counsel), https://www.fcc.gov/news-events/blog/2020/10/21/fccs-authority-interpret-section-230-communications-act. Anyway, this has never been a role that the Commission has assumed for itself. *See* Tony Romm, *FCC Push to Rethink Legal Protections for Tech Giants Marks Major Turn Amid Months of Political Pressure*, Wash. Post (Oct. 21, 2020) (internal quotations omitted), https://www.washingtonpost.com/technology/2020/10/16/fcc-facebook-twitter-section-230/. Notably, the drafters of Section 230 removed a provision that would have granted the FCC authority to regulate the internet. See H.R. REP. NO. 104-223, at 29 (1995); 141 CONG. REC. H8469 (daily ed. Aug. 4, 1995); 141 CONG. REC. H9988 (daily ed. Oct. 12, 1995). If this were not enough to dissuade agency action here, the agency's claim to interpretative authority on this subject would bump right into the US Supreme Court's recent hard turn against deference to agency interpretations of law. *See* Loper Bright Enterprises v. Raimondo, US, 144 S.Ct. 2244 (2024).

139 **"reasonable content moderation" guidelines:** Danielle Keats Citron, *How to Fix Section 230*, 103 Boston U. L. Rev. 713, 750–53 (2023). The textual formulation of this concept could vary. Congress could exclude companies that "encourage" or "induce" unlawful expressive acts as well as those that are "deliberately indifferent" about such third-party activity on their services. Danielle Keats Citron & Mary Anne Franks, *The Internet as a Speech Machine and Other Myths Confounding Section 230 Reform*, 2020 U. Chic. Legal Forum 45, 70–71 (2020).

139 **should focus on restraining litigious plaintiff attorneys:** This proposed exception for government civil enforcement was part of one of the few bipartisan bills that would revise Section 230. See US Senator John Thune, Press Release, *Thune, Schatz Introduce Legislation to Update Section 230, Strengthen Rules, Transparency on Online Content Moderation, Hold Internet Companies Accountable for Moderation Practices* (June 24, 2020), https://www.thune.senate.gov/public/index.cfm/2020/6/thune-schatz-introduce-legislation-to-update-section-230-strengthen-rules-transparency-on-online-content-moderation-hold-internet-companies-accountable-for-moderation-practices.

140 **"privacy violations, cyberstalking, or cyber harassment":** Danielle Keats Citron, *How to Fix Section 230*, 103 Boston U. L. Rev. 713, 753 (2023).

140 **including civil rights laws and rules against unfair or deceptive trade practices:** Olivier Sylvain, *Platform Realism, Information Inequality, and Section 230 Reform*, 131 Yale L. Forum 475, 500 (2022).

140 **hierarchy of harms that leaves other harmful conduct to be addressed another day:** Danielle Keats Citron & Mary Anne Franks, *The Internet as a Speech Machine and Other Myths Confounding Section 230 Reform*, 2020 U. Chic. Legal Forum 45, 69–70 (2020).

140 **in cases in which plaintiffs challenge a company's commercial practices and service designs:** *See generally* Olivier Sylvain, *Intermediary Design Duties*, 50 Conn. L. Rev. 203, 242, 270 (2018); Olivier Sylvain, *Platform Realism, Informational Inequality, and Section 230 Reform*, 131 Yale L. J. F. 475, 497 (2021); Olivier Sylvain, *Discriminatory Designs on User Data*, Knight First Amendment Institute at Columbia University (Apr. 1, 2018). https://knightcolumbia.org/content/discriminatory-designs-user-data.

140 **James Grimmelman has warned:** James Grimmelman, *To Err Is Platform*, Knight First Amendment

Institute at Columbia University (Apr. 6, 2018) (response to Olivier Sylvain's essay "Discriminatory Designs on User Data"), https://knightcolumbia.org/content/err-platform.

141 **against Facebook, Instagram, YouTube, TikTok, and Snapchat:** *See* In re Social Media Adolescent Addiction/Personal Injury Products Liability Litigation, 753 F.Supp.3d 849 (N.D. Cal. 2024); In re Social Media Adolescent Addiction/Personal Injury Products Liability Litigation, 702 F.Supp.3d 809, 833 (N.D. Cal. 2023). There is a similar case in California state court.

141 **that the companies pace and cluster notifications to induce addiction:** In re Social Media Adolescent Addiction/Personal Injury Products Liability Litigation, 702 F.Supp.3d 809, 829 (N.D. Cal. 2023).

145 **ways in which the platforms collect data or distribute and amplify content:** Tara Wright, *The Platform Transparency and Accountability Act: New Legislation Addresses Platform Data Secrecy*, Stanford Cyber Policy Center (Dec. 9, 2021), https://law.stanford.edu/press/the-platform-transparency-and-accountability-act-new-legislation-addresses-platform-data-secrecy/.

145 **alert policymakers to the potential harms that the companies' service features cause:** California's transparency rules appear to have had this effect. *See generally* Electronic Privacy Information Center, Assessing the Assessments (2025) (Mayu Tobin-Miyaji is the principal author), https://epic.org/wp-content/uploads/2025/06/Assessing-the-Assessments-Report.pdf.

145 **most familiar example of this kind of regulation:** *See* 42 US Code § 4332.

146 **European Union's Digital Services Act disclosure and impact assessment requirements:** *See generally* European Commission, *How the Digital Services Act Enhances Transparency Online*, https://digital-strategy.ec.europa.eu/en/policies/dsa-brings-transparency (last visited, Dec. 6, 2024).

146 **reports on their content moderation practices every year:** See generally European Commission, *The Future of European Competitiveness* (Sept. 2024), https://commission.europa.eu/topics/strengthening-european-competitiveness/eu-competitiveness-looking-ahead_en.

146 **large social media and search engines, for example:** European Commission, Press Release, *Digital Services Act: Commission Designates First Set of Very Large Online Platforms and Search Engines* (Apr. 24, 2023),

 https://ec.europa.eu/commission/presscorner/detail/en/ip_23_2413.

146 **also identify and analyze the risk that they distribute unlawful content:** Digital Services Act, Article 34(1). They must also assess whether their services negatively impact "freedom of expression," "civic discourse," and "electoral processes." The DSA is notably specific about the service design features that ought to be part of a company's risk assessment, including those of "recommender systems and any other relevant algorithmic system," "their content moderation systems," and "systems for selecting and presenting advertisements." Digital Services Act, Article 34(2).

146 **the efficacy of their risk mitigation measures:** The Commission will promulgate rules for implementing this researcher access requirement in 2025 after having undertaken a public consultation in late 2024. See European Commission, Press Release, *Commission Launches Public Consultation on the Rules for Researchers to Access Online Platform Data Under the Digital Services Act* (Oct. 29, 2024), https://digital-strategy.ec.europa.eu/en/news/commission-launches-public-consultation-rules-researchers-access-online-platform-data-under-digital.

Columbia Global Reports is a nonprofit publishing imprint from Columbia University that commissions authors to produce works of original thinking and on-site reporting from all over the world, on a wide range of topics. Our books are short—novella-length, and readable in a few hours—but ambitious. They offer new ways of looking at and understanding the major issues of our time. Most readers are curious and busy. Our books are for them.

If this book changed the way you look at the world, and if you would like to support our mission, consider making a gift to Columbia Global Reports to help us share new ideas and stories.

Visit globalreports.columbia.edu to support our upcoming books, subscribe to our newsletter, and learn more about Columbia Global Reports. Thank you for being part of our community of readers and supporters.

The Milk Tea Alliance: Inside Asia's Struggle Against Autocracy and Beijing
Jeffrey Wasserstrom

The Fall of Affirmative Action: Race, the Supreme Court, and the Future of Diversity in America
Justin Driver

Why Live: How Suicide Becomes an Epidemic
Helen C. Epstein

The Web Beneath the Waves: The Fragile Cables That Connect Our World
Samanth Subramanian

The Robe and the Sword: How Buddhist Extremism Is Shaping Modern Asia
Sonia Faleiro

The Triangle of Power: Rebalancing the New World Order
Alexander Stubb